GUIDEBOOK FOR DEMAND AGGREGATION

WAY FORWARD FOR ROOFTOP SOLAR IN INDIA

DECEMBER 2022

ASIAN DEVELOPMENT BANK

Notes:
1.	In this publication, Ministry of New and Renewable Energy, wherever mentioned, refers to the said Union Ministry of the Government on India.
2.	"FY" before a calendar year denotes the year in which the fiscal year ends, e.g., FY2022 ends on 31 March 2022.
3.	ADB recognizes "Bangalore" as Bengaluru.

On the cover: Wider participation is the key to achieving rooftop solar targets. Photo 1 (ID 3562728 © Igorko82 | Dreamstime.com) depicts a row of houses. Photo 2 (ID 210109335 © Martin Malchev | Dreamstime.com) symbolizes a typical solar roof.

CONTENTS

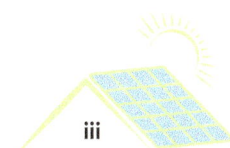

TABLES, FIGURES, AND BOXES

Tables

Figures

Boxes

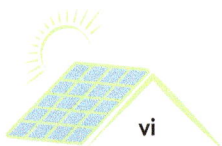

ACKNOWLEDGMENTS

This publication *Guidebook for Demand Aggregation: Way Forward for Rooftop Solar in India* was conceptualized and prepared under the leadership of Jigar Bhatt, Energy Specialist, Asian Development Bank (ADB), India Resident Mission as a part of Technical Assistance (TA) for facilitating Solar Rooftop Investment Program (SRIP) in India. This guidebook has been prepared by KPMG Advisory Services Private Limited and Idam Infrastructure Advisory Private Limited under the ADB SRIP TA. We are indebted to the ADB Management for providing valuable guidance, support, and encouragement for preparing this document. It would not have been possible to present this work in its current form without insightful review and excellent contribution from the agencies and peers—all of which are sincerely appreciated.

We would like to take this opportunity to thank Vandana Kumar, Additional Secretary and Amitesh Kumar Sinha, former Joint Secretary, Ministry of New and Renewable Energy (MNRE), Government of India for motivating us to undertake this study. We would also like to express our sincere gratitude to Jeevan Kumar Jethani, Senior Director, MNRE for not only sharing his time and valuable knowledge but for supporting us through the process of developing and finalizing this publication.

We take this opportunity to thank the senior management of the state government agencies in Kerala, Goa, and Karnataka who have initiated and launched demand aggregation programs in these partner states for increasing the penetration of rooftop solar projects. We also thank SUPRABHA TA Program of the World Bank and Indo-German Energy Program of Deutsche Gesellschaft für Internationale Zusammenarbeit (GIZ) GmbH for sharing valuable comments, best practices, and success stories from the states they support, that is Madhya Pradesh and Gujarat.

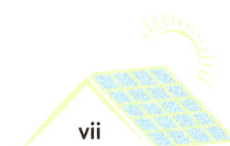

ABBREVIATIONS

ADB	Asian Development Bank
BESCOM	Bangalore Electricity Supply Company
BRPL	BSES Rajdhani Power Limited
C&I	commercial and industrial
CAPEX	capital expenditure
EOI	expression of interest
EPC	engineering, procurement, and construction
GW	gigawatt
IPGCL	Indraprastha Power Generation Company Limited
KSEB	Kerala State Electricity Board
kW	kilowatt
kWh	kilowatt-hour
kWp	kilowatt-peak
MNRE	Ministry of New and Renewable Energy
MW	megawatt
O&M	operation and maintenance
PPA	power purchase agreement
RESCO	Renewable Energy Service Company
RTPV	rooftop photovoltaic
RTS	rooftop solar
SMC	Surat Municipal Corporation
T&D	transmission and distribution
WRI India	World Resources Institute India

EXECUTIVE SUMMARY

By 2022, India was aiming to generate 175 gigawatts (GW) of renewable energy, including 100 GW of solar energy—with 40 GW of that output generated through rooftop solar (RTS) projects. However, despite an enabling policy environment and attractive economics, progress towards that target has been slow. The Government of India has enhanced its renewable energy deployment target to 500 GW by 2030. The total installed capacity of RTS in India by 28 February 2022 was only 6.476 GW, making it imperative to identify suitable business models and develop new market mechanisms to drive RTS adoption. The regulations and framework of the Forum of Regulators stress the need to promote and facilitate new and innovative models for the installation of RTS systems and envisage the role of utilities in their deployment.[1]

The utility can play an active role in aggregating demand and drive the adoption of RTS systems through a facilitation or investment approach. The few utility-led demand aggregation models that have been implemented have shown promising results so far. The overall demand aggregation process can be broadly categorized into three phases—(i) aggregating interest from consumers, (ii) feasibility assessment, and (iii) deployment of RTS projects (Figure E.1).

Figure E.1: Main Stages of the Demand Aggregation Program

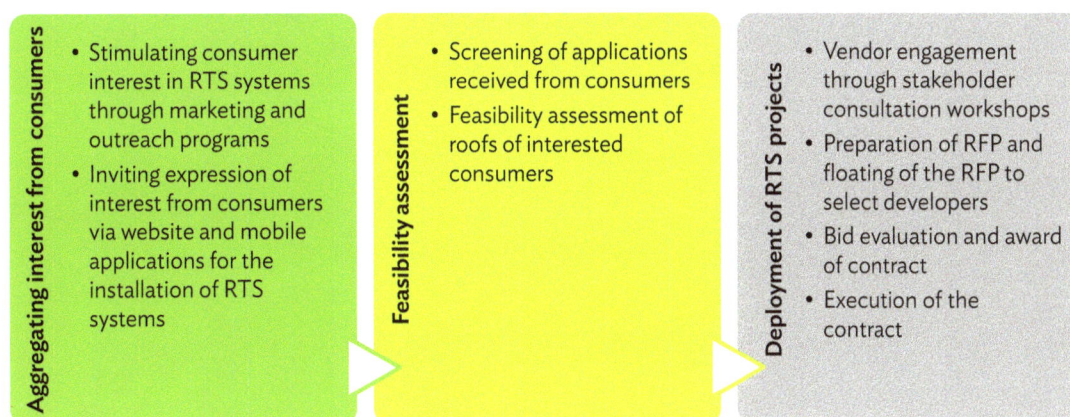

Aggregating interest from consumers
- Stimulating consumer interest in RTS systems through marketing and outreach programs
- Inviting expression of interest from consumers via website and mobile applications for the installation of RTS systems

Feasibility assessment
- Screening of applications received from consumers
- Feasibility assessment of roofs of interested consumers

Deployment of RTS projects
- Vendor engagement through stakeholder consultation workshops
- Preparation of RFP and floating of the RFP to select developers
- Bid evaluation and award of contract
- Execution of the contract

RFP = request for proposal, RTS = rooftop solar
Source: ADB Solar Rooftop Investment Program Technical Assistance.

---

[1] Government of India, Ministry of New and Renewable Energy. 2022. Programme/Scheme wise Cumulative Physical Progress as on Feb, 2022, Physical Progress. https://mnre.gov.in/the-ministry/physical-progress.

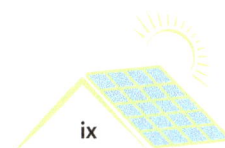

For a utility to choose between acting as a facilitator or an investor in RTS deployment, it must clearly understand the objective of demand aggregation. Under the facilitation role, the utility aggregates interest from consumers and takes the steps necessary to engage vendors to install the RTS system on the interested consumers' roofs. However, in an investment role the utility invests, owns, and operates RTS projects by leasing consumers' roofs on a "right-to-use" basis. The utility can choose between an investment or a facilitation role based on the factors discussed in the guidebook.

Appropriate demand aggregation programs can also help reduce the transactional and consumer acquisition costs which were escalated by the lockdowns and restrictions on travel and business operation hours imposed by the corona-virus disease (COVID-19) pandemic.

Under Phase II of the grid-connected RTS program of the Ministry of New and Renewable Energy (MNRE), utilities are designated as implementing agencies for the deployment of RTS projects, facilitating the disbursal of central financial assistance (CFA) to residential consumers according to prevailing guidelines. Of the allocated CFA, 3% in service charges are payable to the utility for implementation of the program—covering such tasks as (i) aggregating demand; (ii) creating efficient RTS cells in distribution companies; (iii) managing the bidding process; (iv) implementation; (v) creating and streamlining the process of net metering and billing; (vi) inspecting, monitoring, and developing online portals; (vii) training distribution company officials; (viii) delivering awareness programs to proliferate rooftop solar photovoltaic projects; and (ix) creating and operating project management groups in MNRE.

The guidebook will assist distribution companies or utilities interested in undertaking demand aggregation programs in their license areas. It describes the approaches to program implementation and lists the important parameters that any utility should consider before designing a demand aggregation program. It also provides step-by-step guidance on undertaking such a program.

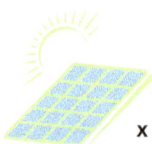

1 INTRODUCTION

Grid-connected renewable electricity generation in India has evolved significantly over the last few years. Emphasis on cleaner energy production and the competitive electricity market embedded in the Electricity Act, 2003 has laid a robust foundation for renewable energy in India. Maturing technology, market consolidation, and substantial cost reduction along with strong policy and regulatory support have resulted in a multifold increase in the share of renewable generation capacity.

India has the fourth-highest share of greenhouse gas emissions (5.7%) in annual global emissions, behind the People's Republic of China (PRC), the United States (US), and the European Union (EU).[2] Consequently, India is under increasing global pressure to reduce its overall carbon emissions. These can be most effectively mitigated by adopting renewable resources to meet energy demand.

By 2022, India was aiming to generate 175 gigawatts (GW) of renewable energy, including 100 GW of solar energy—with 40 GW of that output generated through rooftop solar (RTS) projects.

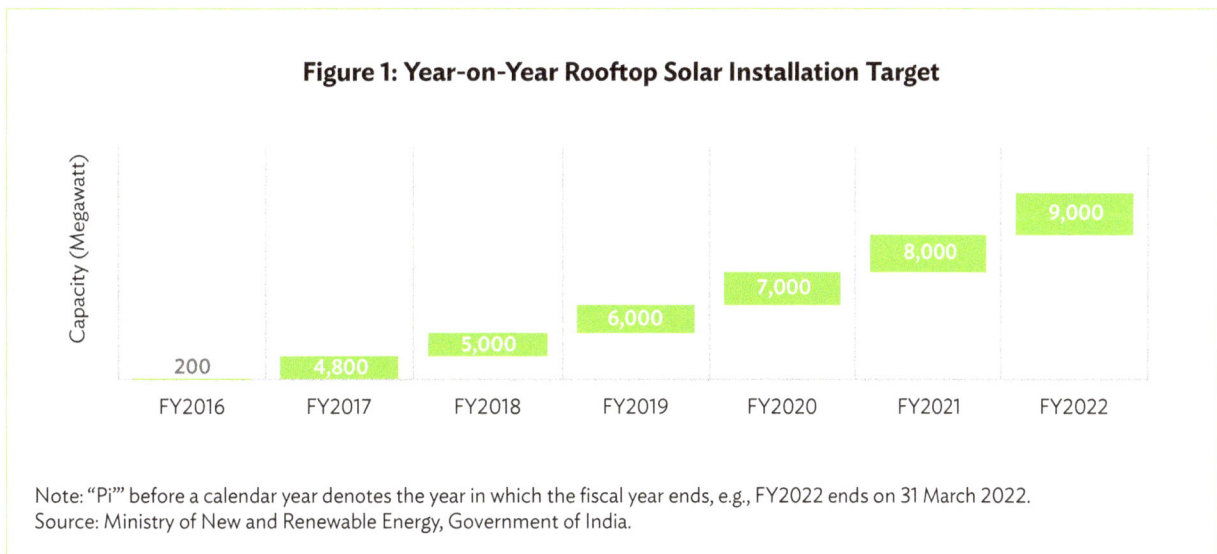

Figure 1: Year-on-Year Rooftop Solar Installation Target

Note: "Pi'" before a calendar year denotes the year in which the fiscal year ends, e.g., FY2022 ends on 31 March 2022.
Source: Ministry of New and Renewable Energy, Government of India.

The Ministry of New and Renewable Energy (MNRE) has also indicated tentative targets in proportion to the state-wise power consumption and consequently solar power requirement to meet the corresponding renewable purchase obligation (RPO) (Figure 2).

---

2   United Nations Environment Programme. 2021. The Emissions Gap Report. UNEP Denmark Technical University Partnership. https://www.unep.org/resources/emissions-gap-report-2021.

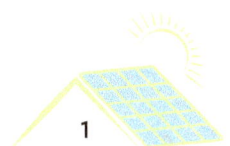

Figure 2: Rooftop Solar Target Allocation by State, 2015 (Gigawatt)

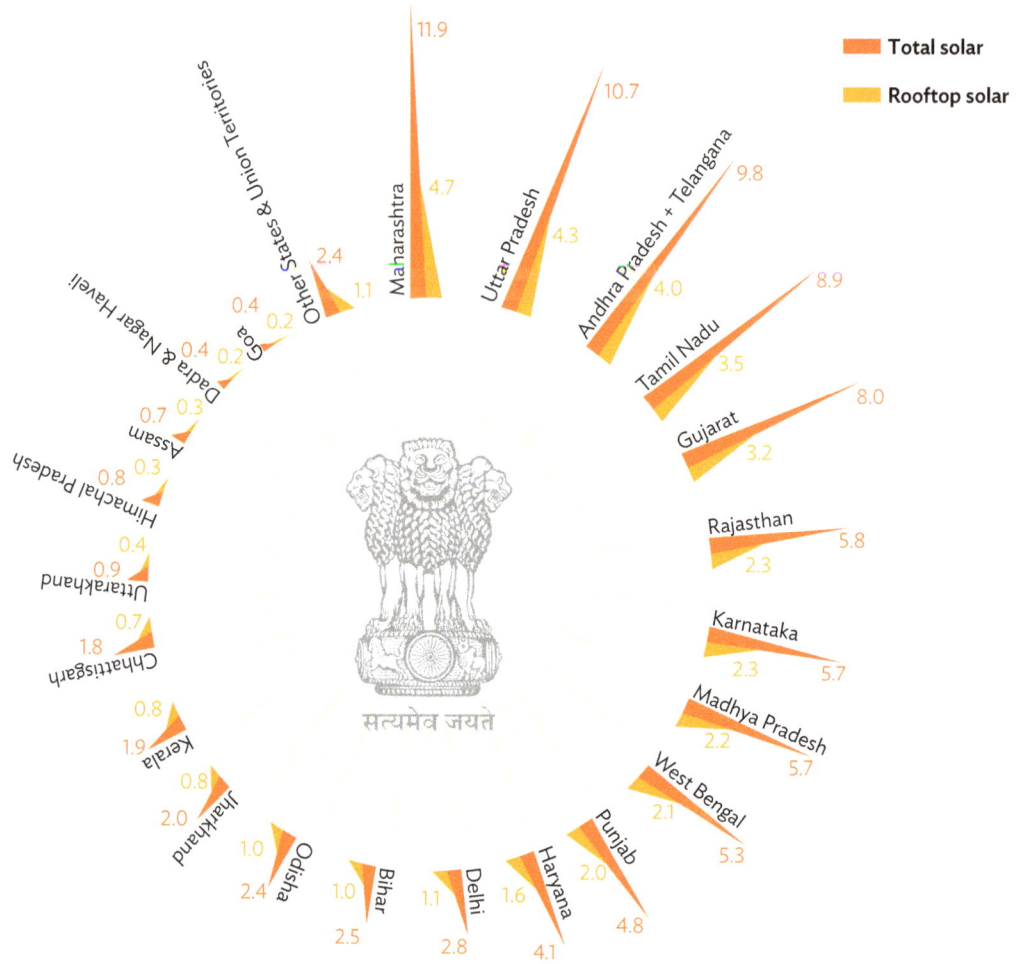

Legend:
- Total solar
- Rooftop solar

Maharashtra 11.9 / 4.7
Uttar Pradesh 10.7 / 4.3
Andhra Pradesh + Telangana 9.8 / 4.0
Tamil Nadu 8.9 / 3.5
Gujarat 8.0 / 3.2
Rajasthan 5.8 / 2.3
Karnataka 5.7 / 2.3
Madhya Pradesh 5.7 / 2.2
West Bengal 5.3 / 2.1
Punjab 4.8 / 2.0
Haryana 4.1 / 1.6
Delhi 2.8 / 1.1
Bihar 2.5 / 1.0
Odisha 2.4 / 1.0
Jharkhand 2.0 / 0.8
Kerala 1.9 / 0.8
Chhattisgarh 1.8 / 0.7
Uttarakhand 0.9 / 0.4
Himachal Pradesh 0.8 / 0.3
Assam 0.7 / 0.3
Dadra & Nagar Haveli 0.4 / 0.2
Goa 0.4 / 0.2
Other States & Union Territories 2.4 / 1.1

Note: Rooftop Solar targets are 2 GW each for Andhra Pradesh and Telangana.
Sources:
(i) Total solar target: Government of India, Ministry of New and Renewable Energy. 2015. Tentative State-wise Break-up of Renewable Power by 2022. http://164.100.94.214/sites/default/files/uploads/Tentative-State-wise-break-up-of-Renewable-Power-by-2022.pdf.
(ii) Rooftop solar target: Government of India, Ministry of New and Renewable Energy. 2015. State-wise and year-wise target for installation of 40,000 MWp GCRT systems. http://164.100.94.214/sites/default/files/webform/notices/State-wise-and-year-wise-target-for-installation-of-40000MWp-GCRT-systems_0.pdf

States have undertaken several policies and regulatory initiatives to promote RTS. These include the provision of subsidies for specific consumer categories and net metering regulations. Most states have introduced dedicated solar policies and net metering regulations to facilitate the scaling up and adoption of grid-connected RTS systems across different consumer categories. The adoption has been highest in the commercial and industrial (C&I) consumer category, largely driven by declining solar installation costs which have led to a decrease in solar tariffs in comparison with utility tariffs for the category. By February 2022, the top five states accounted for around half of the total RTS capacity in India. Gujarat has the highest installed capacity of RTS (1,736 megawatts [MW]) followed by Maharashtra (921 MW), Rajasthan (689 MW), Haryana (397 MW) and Tamil Nadu (333 MW).

Despite an enabling policy environment and attractive economy, the progress in achieving the target has been slow. By 28 February 2022, the total installed capacity of India's RTS system was only 6.48 GW.[3]

---

[3] Government of India, Ministry of New and Renewable Energy. 2022. Programme/Scheme wise Cumulative Physical Progress as on Feb, 2022, Physical Progress. https://mnre.gov.in/the-ministry/physical-progress.

The slow growth is attributed to several factors ranging from lack of access to finance and the disinterest of utilities to difficulty in choosing credible project developers and vendors (Figure 3).

Given the low rate of deployment of RTS so far, it becomes imperative to identify new business models and develop new market mechanisms to drive its adoption. The utility-driven demand aggregation model is one such, wherein the utility aggregates the rooftop owners in its license area and identifies and impanels developers who will install systems in the aggregated rooftop space. Experimented with by various utilities and state nodal agencies, the model has led to considerable cost reduction; increase in consumer awareness and interest in RTS installation; and streamlining of RTS demand aggregation and implementation procedures in the states. This model has several incentives for all the stakeholders involved.

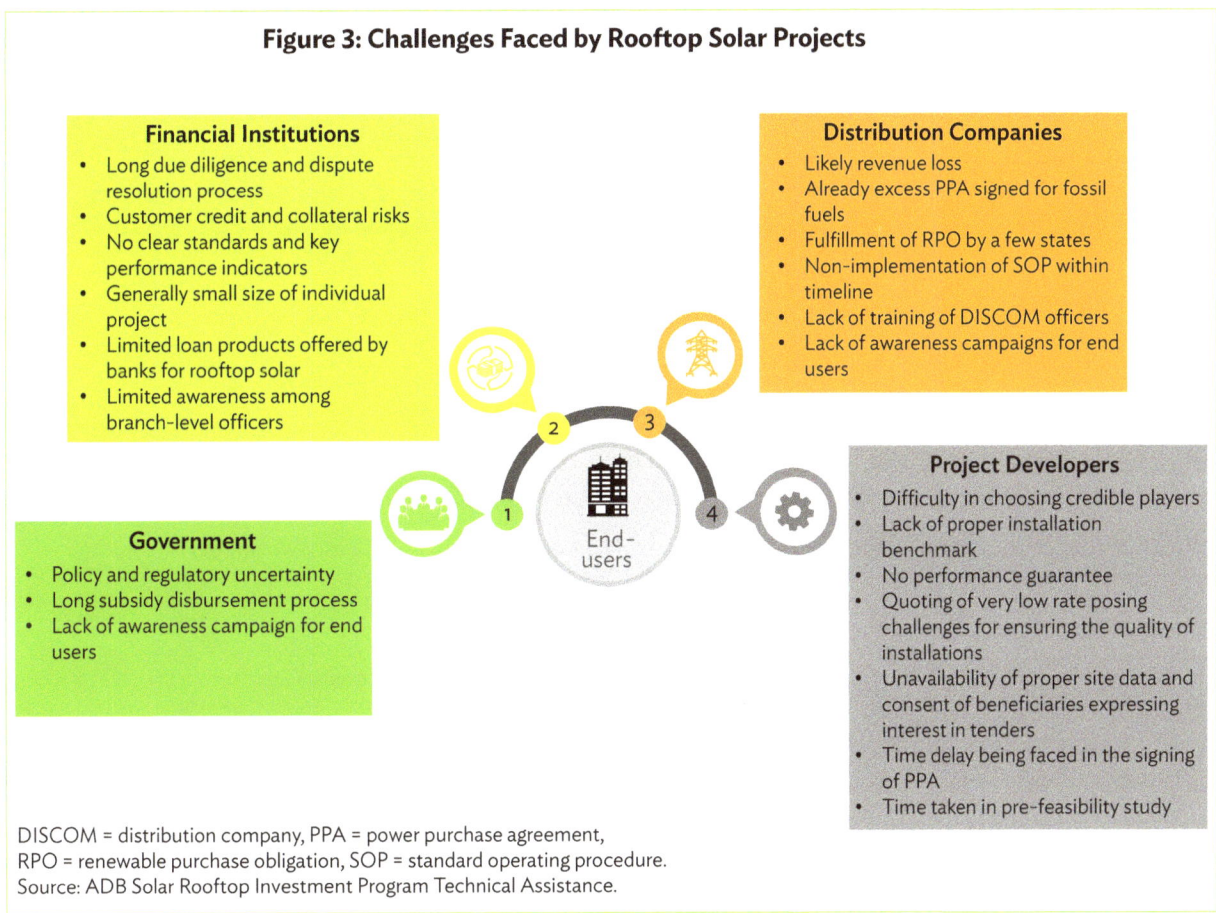

Figure 3: Challenges Faced by Rooftop Solar Projects

Financial Institutions
- Long due diligence and dispute resolution process
- Customer credit and collateral risks
- No clear standards and key performance indicators
- Generally small size of individual project
- Limited loan products offered by banks for rooftop solar
- Limited awareness among branch-level officers

Distribution Companies
- Likely revenue loss
- Already excess PPA signed for fossil fuels
- Fulfillment of RPO by a few states
- Non-implementation of SOP within timeline
- Lack of training of DISCOM officers
- Lack of awareness campaigns for end users

Government
- Policy and regulatory uncertainty
- Long subsidy disbursement process
- Lack of awareness campaign for end users

Project Developers
- Difficulty in choosing credible players
- Lack of proper installation benchmark
- No performance guarantee
- Quoting of very low rate posing challenges for ensuring the quality of installations
- Unavailability of proper site data and consent of beneficiaries expressing interest in tenders
- Time delay being faced in the signing of PPA
- Time taken in pre-feasibility study

End-users

DISCOM = distribution company, PPA = power purchase agreement,
RPO = renewable purchase obligation, SOP = standard operating procedure.
Source: ADB Solar Rooftop Investment Program Technical Assistance.

This guidebook will assist utilities interested in undertaking demand aggregation programs in their license areas by
- introducing solar rooftop targets and demand aggregation programs;
- explaining the need for alternative approaches;
- discussing demand aggregation benefits and emerging utility-led business models;
- examining the existing policy and regulatory framework for solar rooftop sector;
- exploring case studies for similar projects and key learnings from these programs;
- outlining the steps for undertaking demand aggregation program; and
- paving a way forward for interested utilities.

2 THE NEED FOR NEW APPROACHES TO SCALING UP THE DEPLOYMENT OF ROOFTOP SOLAR SYSTEMS

India's current RTS installations have largely been deployed under consumer-owned capital expenditure (CAPEX) or project developer-owned operational expenditure (OPEX) business models with limited outreach. In the **CAPEX model**, the entire investment comes from the power consumer and from central and state governments in the form of capital subsidy (wherever applicable). The consumer hires a solar engineering, procurement, and construction (EPC) company, which provides turnkey installation of the entire solar power system and hands over the assets to the consumer. The company may also take care of annual plant operation and maintenance (O&M) for a limited period (generally 5 years) depending on mutually agreed cost and terms embedded in the annual maintenance contract.

In the **OPEX model,** an investor or project developer as a Renewable Energy Service Company (RESCO) invests in CAPEX and the consumer pays for the energy consumed from the dedicated solar power set up. Both consumer and developer sign a long-term power purchase agreement (PPA) for an agreed tenure (generally 25 years) and tariff.

However, these business models have several limitations (Table 1).

Table 1: Challenges of Conventional Business Models

Capital expenditure model	Operational expenditure model
Limited awareness among consumers	Small individual rooftop size not preferred by the third party
Lack of confidence among consumers and financial institutions in the technology	Clearances, contractual and payment risks
Limited access to finance	Limited access to finance
Loading of operational risks on the consumer who has limited technical knowledge	Restrictions on RESCO models imposed by state governments

RESCO = Renewable Energy Service Company.
Source: ADB Solar Rooftop Investment Program Technical Assistance.

Utilities in India have so far been reluctant to promote RTS systems, apprehensive of the loss of revenue from subsidizing the commercial and industrial (C&I) consumer base. If utilities actively engage in the market-making process, they could facilitate RTS deployment in their preferred consumer categories, creating and leveraging the opportunity to balance the revenue losses due to subsidization. This could reduce both transmission and distribution (T&D) losses and cost of supply to the subsidized consumer category, and at the same time improve revenue streams. The MNRE explicitly recognizes these advantages under Phase II of the grid-connected RTS program while designating utilities as implementing agencies.

Distribution utilities, consumers, rooftop developers, and financial institutions are key stakeholders in the large-scale deployment of RTS. Inherently smaller and more widely dispersed than large-scale solar projects, RTS demands much

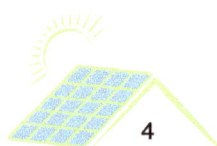

more time and effort from developers to engage consumers. Demand aggregation and a more constructive role of utilities in the RTS sector could expand market opportunities and help in reducing the transaction costs and risks for the developers. It could build confidence and streamline the processes of consumer acquisition, procurement, quality assurance, etc. (Figure 4).

Figure 4: Benefits of Rooftop Solar Installation through the Demand Aggregation Program

State Government
- Reduction in subsidies extended to residential and agricultural consumers and those living below the poverty line
- Reduced reliance on fossil fuel
- Development of local economies and job creation

Utility
- Reduction in transmission and distribution losses
- Cheaper power procurement
- Reduced cost of supply to subsidized consumers
- Investment by utility, increasing revenue and profitability
- Retention of consumers

Project Developers
- Economies of scale on aggregated demand
- Access to cluster of roofs
- Streamlined regulatory approval process
- Easy financing due to bundled nature of projects and assured offtake of electricity generated

Consumers
- Savings on electricity bill
- Zero or minimum capital and operational expenditure
- Assurance on quality of systems
- Avoidance of developer identification process

Financial Institutions
- Reduce transaction cost
- Reduced technology risks due to standard technology deployment
- Time bound implementation
- Appropriate contractual provisions to ensure quality operations and timely maintenance

Source: ADB Solar Rooftop Investment Program Technical Assistance.

The MNRE's Phase II of the grid-connected RTS program has opened up many avenues for adopting a programmatic approach and developing demand aggregation models for RTS systems.[4] Demand aggregation can overcome several challenges associated with conventional RTS deployment and make the process more participatory in nature.

---

[4] A programmatic approach is defined by a set of actions to be undertaken for implementing a project with a long-term aim. A program involves the management of several interdependent activities defined by targets, timelines, plan of action, and roles and responsibilities of stakeholders involved. A programmatic approach ensures consistency throughout the project and allows disparate stakeholders to take a unified view of the ultimate objective and results to be achieved.

In 2009, the world's first community-driven, aggregated RTS project was developed in Portland, US (Box 1).

Box 1: Solarise Portland Community-Driven Program

During the Solarise Portland campaign in 2009, key activities undertaken by the Portland community ranged from generating awareness to informed decision-making (Figure B.1).

Figure B.1: Key Components of the Solarise Portland Campaign

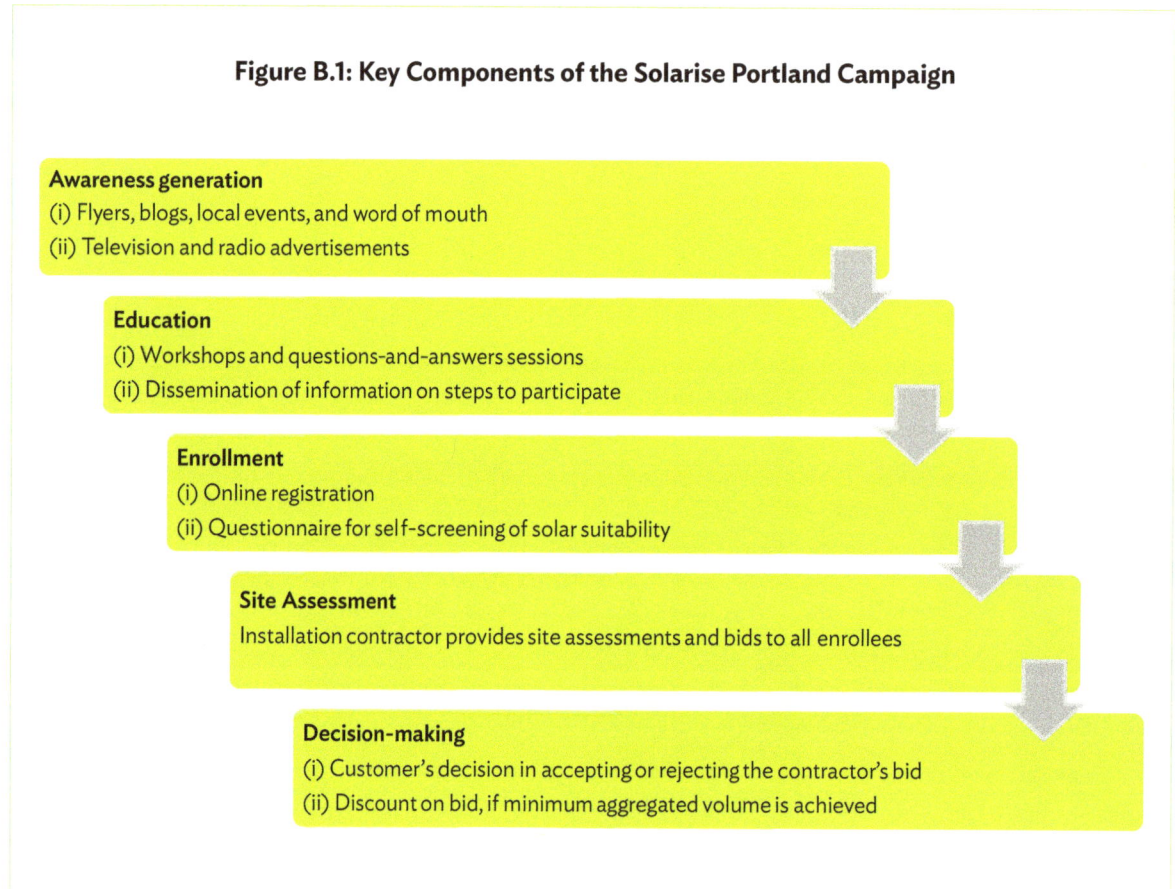

Awareness generation
(i) Flyers, blogs, local events, and word of mouth
(ii) Television and radio advertisements

Education
(i) Workshops and questions-and-answers sessions
(ii) Dissemination of information on steps to participate

Enrollment
(i) Online registration
(ii) Questionnaire for self-screening of solar suitability

Site Assessment
Installation contractor provides site assessments and bids to all enrollees

Decision-making
(i) Customer's decision in accepting or rejecting the contractor's bid
(ii) Discount on bid, if minimum aggregated volume is achieved

continued on next page

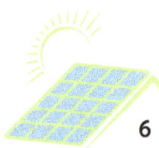

The campaign spearheaded by community-based organizations was supported by technical advisors and industry associations (Figure B.2).

Figure B.2: Key Stakeholders in the Solarise Portland Campaign

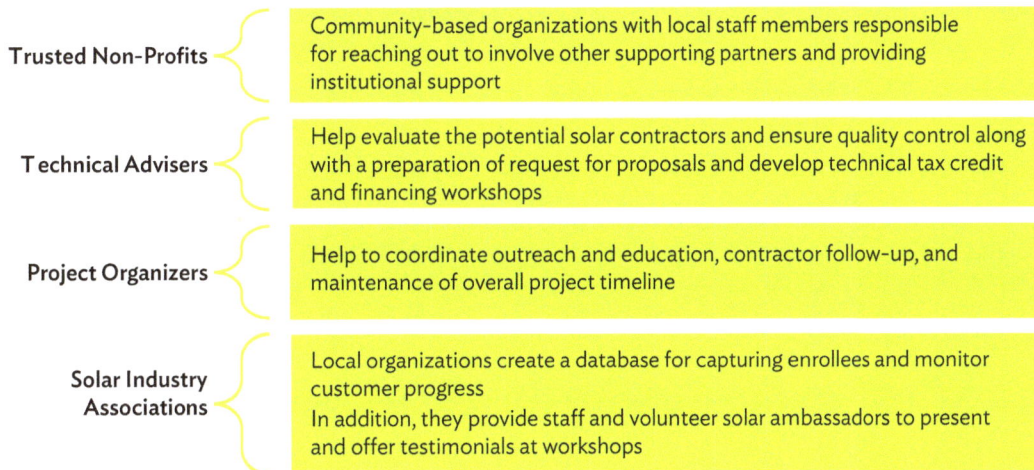

Trusted Non-Profits — Community-based organizations with local staff members responsible for reaching out to involve other supporting partners and providing institutional support

Technical Advisers — Help evaluate the potential solar contractors and ensure quality control along with a preparation of request for proposals and develop technical tax credit and financing workshops

Project Organizers — Help to coordinate outreach and education, contractor follow-up, and maintenance of overall project timeline

Solar Industry Associations — Local organizations create a database for capturing enrollees and monitor customer progress
In addition, they provide staff and volunteer solar ambassadors to present and offer testimonials at workshops

The Solarise Portland campaign during 2009–2011 transformed the solar power market. Rooftop solar was installed in 560 homes in Portland with the aggregated capacity of 1.7 MW, driving down market prices by more than 30% across the board and generating over 50 permanent green jobs for site assessors, engineers, project managers, journeyman electricians, and roofers. In 2011, the market was revolutionized with the entry of solar lease and prepaid power purchase agreement options, which have subsequently been used in over half the new solar installations.

Solarise Portland was replicated later at multiple locations. In Solarise Massachusetts, a total of 46 campaigns were organized between 2011 and 2013, through which 2,448 RTS projects were implemented, aggregating to approximately 16 MW capacity. The average cost of installation was $4.15 per watt in Phase I, which reduced to $4 per watt in Phase II.

3 UPCOMING DEMAND AGGREGATION AND UTILITY-DRIVEN BUSINESS MODELS

A. Demand Aggregation Models

Demand aggregation may broadly be enabled through either the facilitation approach or the investment approach (Table 2). Under the former, for a facilitation charge, the utility aggregates projects and facilitates the procurement of systems and solar power services for the consumer. In the investment approach, besides aggregating demand, the utility also invests in developing the projects. The energy generated is either procured by the utility at zero cost or sold to the respective consumers at a predetermined tariff.

Table 2: Approaches to Utility-Led Demand Aggregation

Phase of implementation	Facilitation approach	Investment approach
Procurement	The utility	The utility
	(i) aggregates available roofs from consumers and conducts initial site surveys,	(i) aggregates roofs from consumers and conducts the initial site survey,
	(ii) conducts bidding for aggregated capacity,	(ii) conducts bidding for aggregated capacity,
	(iii) develops bidding documents and contracts,	(iii) conducts contracting with consumer for utilizing roof space,
	(iv) invites bids,	(iv) develops bidding documents and contracts,
	(v) assesses and selects the bidder/s from the received bids, and	(v) invites bids,
	(vi) facilitates the signing of standardized contracts	(vi) assesses and selects the bidder/s from the received bids, and
		(vii) conducts contracting with the developer.
Execution	The utility	The utility
	(i) supervises progress and	(i) arranges financing for investment requirement and
	(ii) conducts regular quality checks.	(ii) builds plants through contracted engineering, procurement, and construction company.
Operation	(i) The utility supervises operation and maintenance (O&M).	(i) The utility conducts O&M (a back-to-back O&M contract can be arranged).

Source: ADB Solar Rooftop Investment Program Technical Assistance.

Additional roles for the utility under the facilitation approach could include facilitating debt financing and collection of monthly installments from consumers, which would provide security for banks to finance small rooftop projects. Otherwise, the utility could also collect payment from consumers for the consumption of electricity from the RTS

plants operated under RESCO mode. This would provide higher payment security to the developers. Table 3 presents the benefit to stakeholders under both approaches.

Table 3: Benefits to Stakeholders in Utility-Driven Demand Aggregation Models

Stakeholder	Facilitation approach	Investment approach
Consumer	(i) Reduction in the cost of system and services (ii) Better performance due to distribution company (DISCOM)-led supervision (iii) Reduced risks of default on services from the engineering, procurement, and construction (EPC) developer or the renewable energy service company (RESCO)	(i) Income from leasing out roof space
EPC Developer or RESCO	(i) Attractive project sizing (ii) Reduced customer acquisition cost (iii) Lower cost of procurement (iv) Smooth net metering/ gross metering connection arrangement	(i) Reduced customer acquisition cost (ii) Attractive project sizing (iii) Lower cost of procurement/sale (iv) Higher payment security (v) Smooth net metering/ gross metering connection arrangement
DISCOM	(i) Income from facilitation fee (ii) Maximization of utility benefits like reduced cross-subsidy burden, reduced distribution transformer congestion, minimized revenue loss, and reduced marginal power purchase cost	(i) Earnings from project returns (ii) Higher control over assets (iii) Maximization of utility benefits like reduced cross-subsidy burden, reduced distribution transformer congestion, minimized revenue loss, and reduced marginal power purchase cost
Financial Institutions	(i) Reduced transaction cost (ii) Reduced technology risks	(i) Reduced transaction cost (ii) Reduced technology and operational risks (iii) Assured offtake (iv) Appropriate payment and credit guarantee mechanisms

Source: ADB Solar Rooftop Investment Program Technical Assistance.

Under these approaches, utilities are responsible for identification of the consumer category and needs, identification of the appropriate business model, and the development of a financial package. The utility can identify the consumer category based on the assessment of T&D losses or revenue realization. Further, the utility can conduct bulk procurement for aggregated roofs allowing a reduction in capital costs. Accordingly, the utility can develop appropriate business models.

B. Utility-Driven Business Models for Rooftop Solar

The Forum of Regulators in its report on "Metering Regulations and Accounting Framework for Grid Connected Rooftop Solar PV in India," dated April 2019, has suggested the following business models:

1. Consumer-owned model (utility only aggregates).
2. Consumer-owned model (utility aggregates and acts as the EPC player).
3. Third-party owned model (utility aggregates and acts a trader between RESCO and the consumer).
4. Utility-owned model (utility aggregates and acts as RESCO).

1. Consumer-Owned Model (Utility Only Aggregates)

In this model, the utility acts as the aggregator by identifying the RTS demand in its distribution circle (Figure 5). The consumers interested in installing RTS systems contact the utility. After the demand aggregation, the utility initiates reverse bidding to provide EPC services to the aggregated demand. The selected EPC service providers sign EPC contracts with interested consumers. The bidders pay a facilitation fee to the utility for the demand aggregation. The utility signs a project management service agreement with the consumers for monitoring the project till interconnection with the grid. The consumer bears the capital expenditure.

Figure 5: Schematic of the Consumer-Owned Model (Utility Only Aggregates)

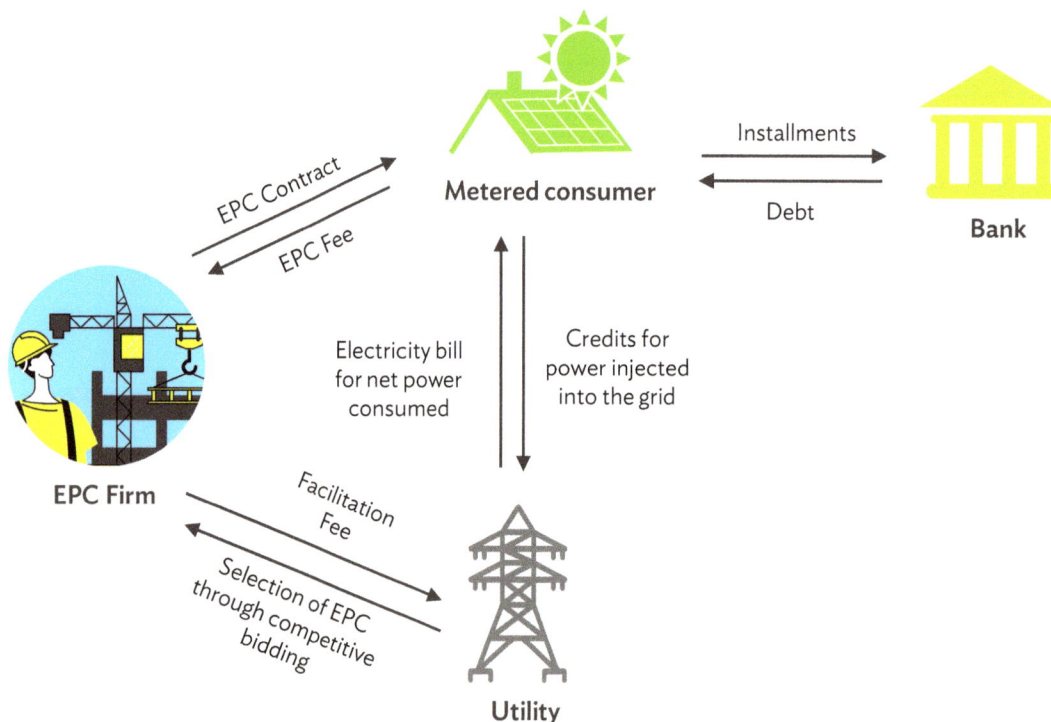

EPC = engineering, procurement, and construction.
Source: ADB Solar Rooftop Investment Program Technical Assistance.

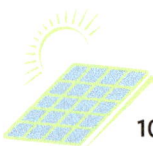

2. Consumer-Owned Model (Utility Aggregates and Acts as Engineering, Procurement, and Construction Player)

In this model, the consumer and the utility sign an EPC contract for the RTS installation (Figure 6). The utility signs another agreement with successful EPC players selected through the reverse bidding. Consumer pays the utility for the EPC services. The utility transfers the same to the EPC firm after charging a margin. In this case, the utility earns revenue from a one-time facilitation fee and a margin on back-to-back EPC agreements.

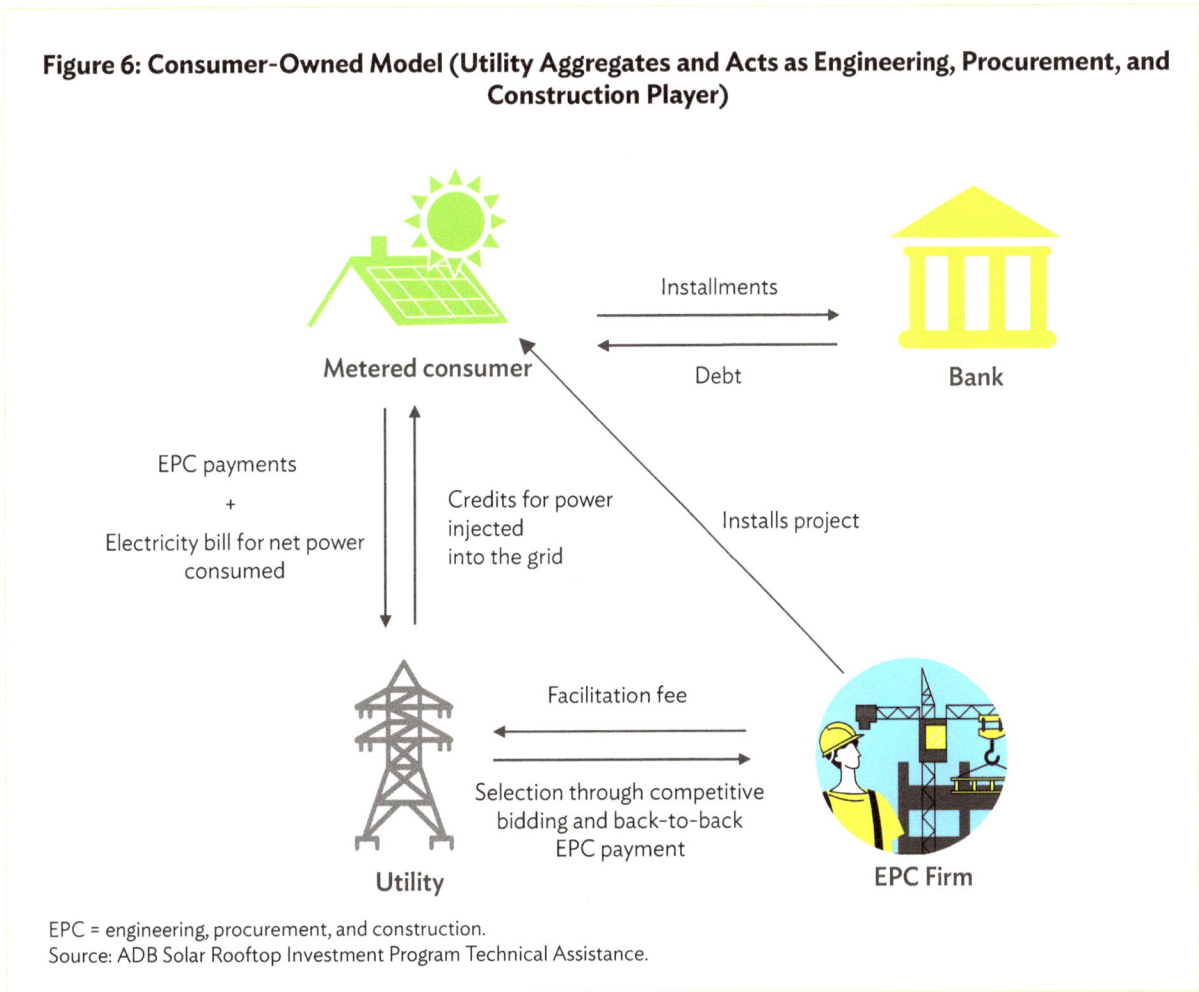

Figure 6: Consumer-Owned Model (Utility Aggregates and Acts as Engineering, Procurement, and Construction Player)

EPC = engineering, procurement, and construction.
Source: ADB Solar Rooftop Investment Program Technical Assistance.

3. Third-Party Owned Model (Utility Aggregates and Acts as the Trader between the Renewable Energy Service Company and the Consumer)

In this model, the utility aggregates the demand in its distribution circle but unlike the models mentioned already, the RESCO is selected based on the reverse bidding (Figure 7). The payment is routed through the utility, which signs the

PPAs with the consumer and the RESCO. The RESCO and the utility sign the PPA, and the utility purchases the power generated from the RTS plant. The utility and the consumer sign a power supply agreement (PSA) for the energy generated by RESCO. The utility charges a trading margin for facilitating the trading options.

Figure 7: Schematic of Third-Party Owned Business Model

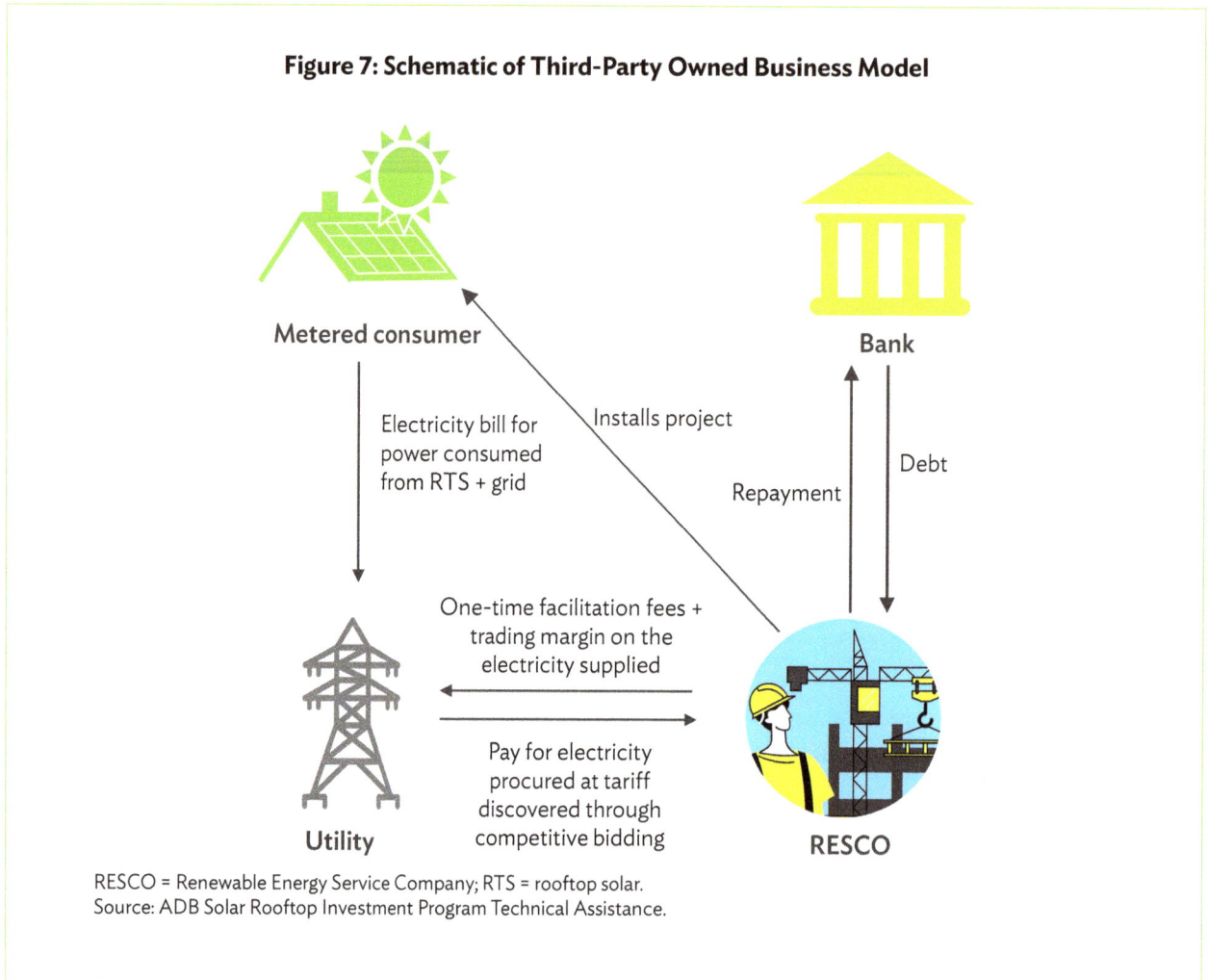

Metered consumer

Bank

Electricity bill for power consumed from RTS + grid

Installs project

Repayment

Debt

One-time facilitation fees + trading margin on the electricity supplied

Pay for electricity procured at tariff discovered through competitive bidding

Utility

RESCO

RESCO = Renewable Energy Service Company; RTS = rooftop solar.
Source: ADB Solar Rooftop Investment Program Technical Assistance.

4. Utility-Owned Model (Utility Aggregates and Acts as RESCO)

In this model, the utility aggregates the demand for RTS systems in its distribution circle and acts as the RESCO by installing the systems on the consumers' roofs (Figure 8). The utility sets up, owns, and operates the RTS systems.

A PPA is signed between the consumers and the utility and EPC contracts are signed between the utility and the EPC firms.

Figure 8: Schematic of Utility-Owned Business Model

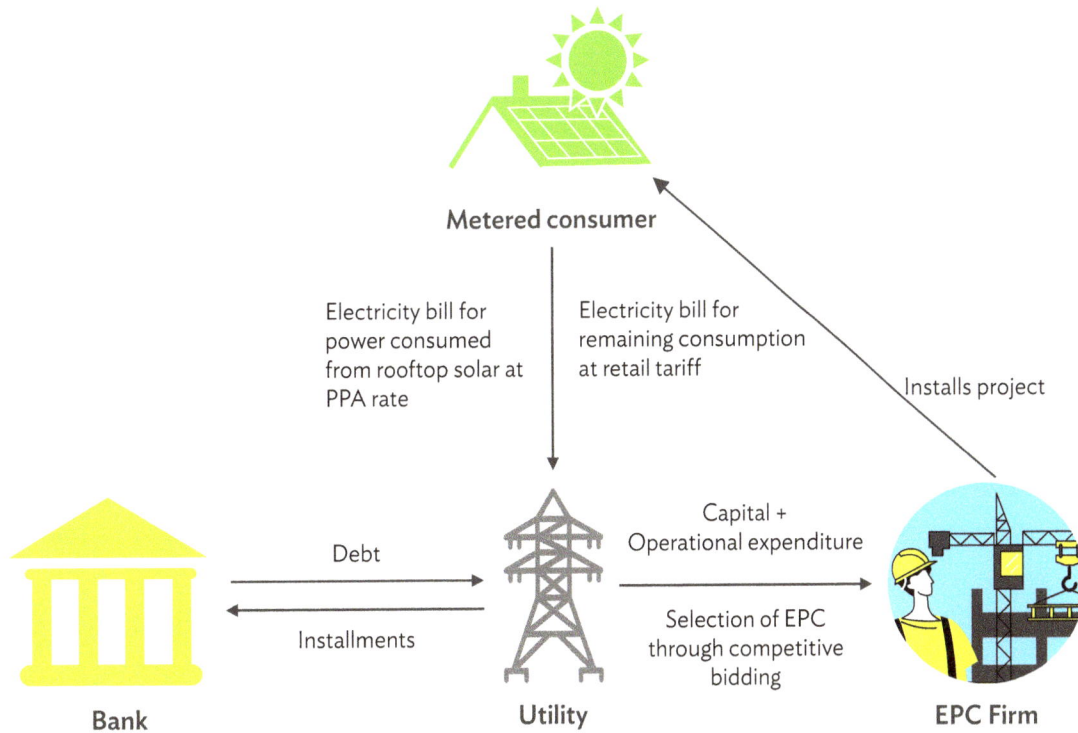

Metered consumer

Electricity bill for power consumed from rooftop solar at PPA rate

Electricity bill for remaining consumption at retail tariff

Installs project

Bank

Debt

Installments

Utility

Capital + Operational expenditure

Selection of EPC through competitive bidding

EPC Firm

EPC = engineering, procurement, and construction; PPA = power purchase agreement. Source: ADB Solar Rooftop Investment Program Technical Assistance.

A. Grid-Connected Rooftop Solar Program Phase II Guidelines

The MNRE has issued Phase II guidelines for promoting the installation of RTS systems connected to the grid with a total financial outlay of ₹118.14 billion. During Phase I, the scheme faced several hurdles, such as involvement of multiple stakeholders (for instance, the state nodal agency, distribution company, project developer, consumer, and the Chief Electrical Inspector to Government), delay in tendering, lack of single window clearance portal, and the lack of awareness among prospective beneficiaries. Reforms introduced through Phase II guidelines in order to streamline the processes and ensure uniformity in implementation are listed below.

(i) Utilities have been identified as the primary implementation agency to reduce multiplicity of stakeholders.

(ii) Single window clearance portal has been developed by the distribution company.

(iii) Central financial assistance (CFA) of up to 40% (earlier 30%) is being made available only for the residential sector for promoting grid-connected RTS systems among such consumers.

(iv) Sustainable business models are being promoted.

(v) Emphasis has been laid on awareness generation, capacity building, human resources development, etc.

(vi) Domestic manufacturing of solar cells and modules is being promoted.

(vii) The customer needs to make the investment toward the cost of the system after deducting the CFA amount.

The scheme primarily has two components as described in Table 4.

Table 4: Components of the Phase II Rooftop Solar Program

Component A: Setting Up of 4 gigawatt (GW) of grid-connected rooftop solar (RTS) projects in the residential sector with central financial assistance (CFA)	
Residential category	**Central financial assistance**
Rooftop solar capacity up to 3 kilowatt (kW)	40%
Rooftop solar capacity above 3 kW and up to 10 kW[a]	40% up to 3 kW Plus 20% for systems above 3 kW and up to 10 kW
Group Housing Societies/Residents Welfare Associations (GHS/RWA) for common facilities up to 500 kilowatt peak (kWp) (@10 kW per house), with the upper limit being inclusive of RTS already installed in GHS/RWA	20%

continued on next page

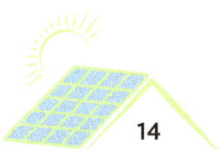

Table 4: continued

Component B: Incentives to utilities based on progress toward achieving 18 GW of grid-connected RTS plants	
Parameters	Incentives to be provided
For installed capacity achieved above 10% and up to 15% over and above of the installed base capacity[b] within a financial year	5% of the applicable cost[c] for the capacity achieved above 10% of the installed base capacity
For installed capacity achieved beyond 15% over and above of the installed base capacity within one financial year	5% of the applicable cost for capacity achieved above 10% and up to 15% of the installed base capacity plus 10% the applicable cost for capacity achieved beyond 15% of the installed base capacity

[a] The residential users may install RTS plants of even higher capacity as provisioned by respective state electricity regulations; however, the CFA will be limited up to 10 kWp capacity.

[b] Installed base capacity shall mean the cumulative RTS capacity installed within the jurisdiction of distribution company at the end of the previous financial year. This will include total RTS capacity installed under residential, institutional, social, government, public sector units, statutory or autonomous bodies, private commercial, industrial sectors, etc.

[c] For calculating incentives for a particular year, the lowest among the rates for RTS discovered by the states or union territories in that year is compared to the lowest benchmark cost of RTS fixed by the MNRE for that year, and the lower of the two is considered as the applicable cost.

Source: Guidelines on implementation of Phase II of the grid-connected rooftop solar program of the Ministry of New and Renewable Energy, Government of India and subsequent amendments.

The primary stakeholders are the MNRE, the utilities, and the vendors impaneled for the supply and installation of the RTS projects (Table 5).

Table 5: Roles and Responsibilities of Various Stakeholders under Phase II

Ministry of New and Renewable Energy	Utilities	Impaneled vendors
Allocate capacity for the ensuing year to different utilities based on the demand and capacity required.	Create rooftop solar (RTS) cell at each division level headed by the executive engineer with the respective sub-divisional officer acting as nodal officer.	Install RTS plant within the time frame decided by the utilities.
Release funds to the utilities for the disbursement of central financial assistance (CFA) to the vendors installing the RTS plant in the residential sector.	Develop a dedicated online portal for grid-connected RTS projects which will be integrated with online application for the Solar Photovoltaic Installation (SPIN) portal.	Use only indigenously manufactured photovoltaic panels (both cells and modules) for projects covered under CFA.
Release eligible incentive to the utilities based on their performance in the last financial year.	Impanel agencies for design, supply, installation, testing, and commissioning of RTS systems on residential premises.	Establish a service center in each district.
Undertake sample physical inspection of the installed RTS plants for which CFA or incentive has been claimed by the utilities.	Notify time-bound procedure for the implementation of the program.	Ensure that service centers cater to the RTS owners within the timelines decided by the utilities, free of cost for the first 5 years (warranty period) of commissioning of the RTS.
Provide necessary assistance to utilities in respect of portal development and its integration with SPIN, capacity building of officials, updating of billing software, etc.	Notify cost of metering arrangements, related connectivity components, and other charges.	

continued on next page

Table 5: continued

Ministry of New and Renewable Energy	Utilities	Impaneled vendors
Undertake capacity building and public awareness campaigns through print and electronic media.	Submit the cumulative capacity of grid-connected RTS plants (in megawatt-peak) installed in their distribution area on 31 March every year for calculating the applicable incentive payable to them by the Ministry of New and Renewable Energy. Ensure that the loss of solar power is avoided due to any reason including grid failure during daytime.	

Source: Guidelines on implementation of Phase II of the grid-connected rooftop solar program of the Ministry of New and Renewable Energy, Government of India.

With clear responsibilities and roles identified for utilities, Phase II expects greater deployment of RTS systems across the states.

B. Rooftop Solar Programs of States

States have undertaken several initiatives for the promotion of RTS plants, such as generation-based incentives, offering of additional subsidies, and demand aggregation for the collective deployment of RTS. Some of the programs are explained in detail in the sections that follow.

1. Delhi Solar Policy, 2016 and Mukhya Mantri Solar Power Yojna (Chief Minister's Solar Power Program)

   (i)   The Government of Delhi provided a generation-based incentive (GBI) of ₹2 per unit on solar generation for a 3-year period between FY2017 and FY2019, as per Delhi Solar Policy 2016 notified vide No. 205 dated 28 September 2015. Government of Delhi has extended the GBI for a period of 5 years starting FY2020.

   (ii) The state nodal agency of Delhi, i.e., the Energy Efficiency and Renewable Energy Management Centre implemented the program through Indraprastha Power Generation Company Limited (IPGCL).

   (iii) The program was applicable to the existing and future net metered connections in the domestic or residential segment only.

   (iv) The minimum eligibility criterion for GBI was 1,100 kilowatt-hour (kWh) solar energy units generated per annum per kilowatt-peak (kWp) and capped up to 1,500 kWh per annum per kWp.

Table 6: Roles and Responsibilities of Stakeholders as per Delhi Solar Policy, 2016

Energy Efficiency and Renewable Energy Management Centre	Utilities	Indraprastha Power Generation Company Limited
Announce solar policy, amendments, and related schemes.	Support installation of solar power plants, their connectivity with the grid network, and metering.	Aggregate the capacity (government rooftops plus the rooftops of "nongovernment" buildings that have a capacity above 50 kW under MNRE or state scheme plus ground-mounted capacities under MNRE or state schemes), float tender, and manage the entire bidding process under both RESCO and CAPEX model.

continued on next page

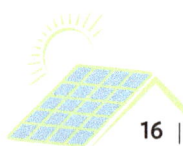

Table 6: continued

Energy Efficiency and Renewable Energy Management Centre	Utilities	Indraprastha Power Generation Company Limited
Allot solar power capacities under various schemes of state and central government and its identified agencies.	Comply with the regulatory framework specified by the Delhi Electricity Regulatory Commission and provisions contained in this policy.	Inspect and assess the potential of RTS as input for the tender.
Provide support in establishing protocols and procedures for easy adoption of solar power.	Adjust the GBI against electricity bill of the eligible consumer.	Perform technical and commercial evaluation of the bid, select the most suitable bidder, and ensure their compliance with technical standards.
Maintain a website for consumers interested in RTS.	Promote online applications for net metering.	Facilitate the signing of the power purchase agreement between the consumer and the winning bidder.
Assist project developers in identifying the technically feasible sites and roofs under the jurisdiction of the state government for the deployment of RTS plants.	Update the status of solar capacity installation with respect to distribution transformers on the website.	Monitor and supervise the timely completion of the RTS project.
Manage the green fund and disbursement of the GBI.		
Support the availing of MNRE subsidies.		
Participate in capacity building and awareness creation.		

CAPEX = capital expenditure, GBI = generation-based incentive, MNRE = Ministry of New and Renewable Energy, OPEX = operational expenditure, RESCO = Renewable Energy Service Company, RTS = rooftop solar.
Source: Government of Delhi. 2016. Delhi Solar Policy, 2016.

In 2018 and 2019, IPGCL aggregated demand for RTS from government buildings, domestic consumers, and housing societies.

The IPGCL invited expression of interest from domestic consumers and housing societies for the implementation of RTS under CAPEX and OPEX models. It initiated the feasibility assessment of the roofs and issued a tender with an aggregated capacity of 40 MW in domestic consumers and housing societies, out of which 35 MW was based on pre-identified buildings and 5 MW was to be implemented in the open category. In the open category, the selected developer was responsible for identifying the buildings and implementing RTS projects at a price discovered through the bidding process. The IPGCL was responsible for monitoring and supervising the installation and commissioning of RTS projects. Further, the IPGCL in recent years has undertaken multiple demand aggregation tenders for residential and government sector consumers. Status of tenders issued by IPGCL through demand aggregation targeting different consumer categories is detailed in Table 7.

Table 7: Tenders Issued by Indraprastha Power Generation Company Limited through Demand Aggregation

Name of Tendering Agency	Tender Capacity and Date of RFP or Tender Publication	Targeted Consumer Category (Data Room referred to in Tender)	Data Room Created / Number of Sites	Actual Implementation Status
IPGCL (MNRE Phase I Scheme)	21.5 MW (published on 13 June 2019)	Government	Yes (18.5 MW / 264 sites)	16.5 MW
IPGCL (MNRE Phase I Scheme)	35 MW (tendered category) + 5 MW (open category) (published on 10 December 2018)	Residential	Yes (12.5 MW / 172 sites)	15.5 MW
IPGCL for BYPL, BRPL, and TPDDL (MNRE Phase II Scheme)	24 MW (tendered category) + 6 MW (open category) (published on 28 February 2020)	Residential	Yes (5.7 MW / 201 sites)	2.119 MW

BRPL = BSES Rajdhani Power Limited, BYPL = BSES Yamuna Power Limited, IPGCL = Indraprastha Power Generation Co. Ltd, MNRE = Ministry of New and Renewable Energy, MW = mega watt, RFP = request for proposal, TPDDL = Tata Power Delhi Distribution Ltd.

Source: Study team analysis based on data drawn from IPGCL.

2. Additional Subsidy by the Government of Goa

The Government of Goa issued an amendment to the Goa State Solar Policy 2017 on 7 February 2019 and designated the Goa Energy Development Agency (GEDA) as a nodal agency. According to the amended policy, the Government of Goa is offering subsidy of 50% (central share 40% and state share 10% for 1–3 kWp RTS system) of the capital cost or the benchmark cost provided by the MNRE or the cost derived through tendering process by GEDA, whichever is lower. The CFA is only available for RTS systems to be implemented in the residential sector and the residential welfare association. Further, the state subsidy is to be released upon completion of 6 months of the solar power being injected into the grid.

C. Model Regulations and Framework for Grid Interactive Distributed Renewable Energy Sources: Forum of Regulators

Regulations and framework of the Forum of Regulators stress on the need to promote and facilitate new and innovative models for the installation of RTS systems and envisage a greater role of the utilities in the deployment of RTS systems. It specifically identifies the roles and responsibilities for the utility.

(i) The utility should publish information on distribution transformer capacity available for adding additional capacity and cumulative installed capacity on its website.

(ii) The utility should undertake technical studies to assess the impact of penetration of distributed renewable energy (DRE) systems on the distribution system.

(iii) The utility shall set up a DRE cell.

(iv) The utility may explore appropriate utility-driven business models, such as demand aggregation, RESCO, and EPC, to promote the installation of DRE in its area of supply.

Enabling policies, incentives, and greater role envisaged for the utilities are expected to drive the deployment of RTS projects to meet the targets set up by the government.

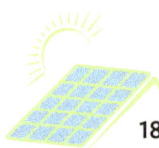

5 EARLY DEMAND AGGREGATION PROGRAMS IN INDIA

The first RTS demand aggregation program was launched in Gandhinagar by the Government of Gujarat with assistance from the International Finance Corporation in public–private partnership (PPP) model in 2010.[5] Of the total aggregated capacity of 5 MW, 80% was deployed in pre-identified government buildings and the remaining in private buildings identified by developers. As the project implementer, the Gujarat Power Corporation Ltd. (GPCL) identified government buildings, while the requisite mix between the private residences and commercial buildings was specified by the government. Being a facilitator, the GPCL initially identified the private rooftop owners who had expressed interest in the program and collected the necessary details (Figure 9). The selected developer subsequently scrutinized that data and selected the potential (technically and legally viable) private owners.

The build–own–operate model was used to set up RTS projects and the power generated from the solar plants was sold to the utility. Minimum price discovered was ₹11.21 per kWh for the implementation of 2.5 MW project (Table 8).

Table 8: Quoted Tariff by Selected Bidders for 2.5 Megawatt Capacity

Developer	Tariff quoted (₹ per kilowatt-hour)
Azure	11.21
Sun Edison	11.793

Source: International Finance Corporation, PricewaterhouseCoopers

The Gandhinagar pilot project was replicated to five more cities of Gujarat with an aggregated capacity of 25 MW. The governments of Madhya Pradesh and Odisha also deployed RTS under the PPP model with an aggregated capacity of 5 MW and 4 MW respectively. Since then, there have been multiple programs for RTS.

Increased volumes of RTS deployed under various programs have

   (i)   lowered transaction costs and improved operations;

   (ii)   built confidence among the stakeholders;

   (iii)   streamlined processes—customer acquisition, procurement, quality systems, etc.;

   (iv)   standardized RTS systems and components;

   (v)   improved the availability of funds from financial institutions; and

   (vi)   accelerated growth in the rooftop sector.

---

[5] International Finance Corporation. 2013. Public-Private Solar Project Generates Power, Reduces Carbon Emissions in India. Climate Change: Stories of Impact. IFC Advisory Services. June. https://www.ifc.org/wps/wcm/connect/9922033d-e6ca-4bab-ade9-11a8fb200cd8/sba-proj-gujaratsolar.pdf?MOD=AJPERES&CVID=jZjHg9V.

Figure 9: Gujarat Demand Aggregation Model

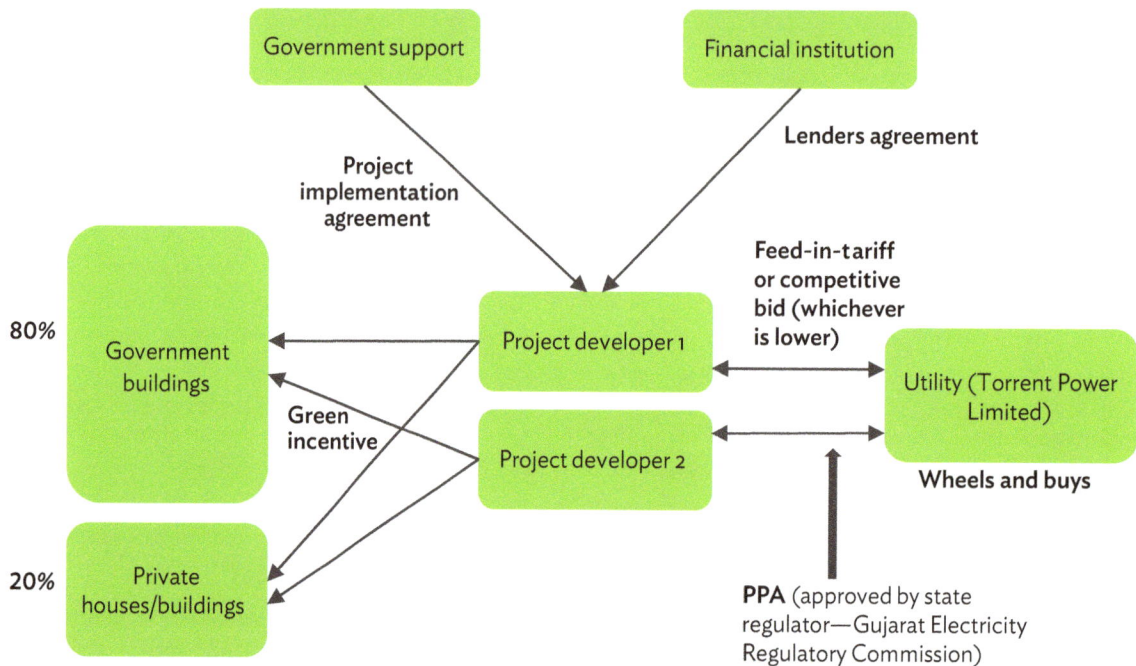

PPA = power purchase agreement.
Source: ADB Solar Rooftop Investment Program Technical Assistance, USAID PACE-D Technical Assistance Program.

In India, the few demand aggregation programs implemented initially resulted in the reduction of overall tariff for solar power. Based on the role played by the aggregator, the programs discussed here are categorized under the facilitation and the investment approach.

A. Facilitation Approach

1. Rooftop Solar Power in Surat Smart City

The Surat Municipal Corporation (SMC) undertook a campaign to popularize RTS plants in Surat Smart City area.[6] It required intensively coordinated effort between the SMC and the residents within a short period of time. Under the smart city initiative, efforts for the deployment of RTS were initiated in September 2016. The SMC acted as the facilitating agency and led the development of the solar guidebook and android application to present a single platform to the consumers. It also conducted extensive promotional campaigns with the help of media and student volunteers.

The project involved the roadmap preparation, stakeholder consultation and cooperation, demand aggregation of the interested consumers, and setting up the interactions with those consumers.

---

[6] A. Datta. 2018. It takes a village – Surat's united effort to embrace solar power. https://www.teriin.org/interview/it-takes-village-surats-united-effort-embrace-solar-power.

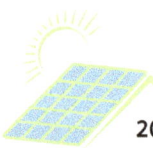

The key activities undertaken in the program are listed below.

(i) **Awareness generation.**

    (a) A campaign to popularize RTS was coordinated.

    (b) Radio, public hoardings, and movie theaters were used.

    (c) College students involved for wider connect.

(ii) **Mobile application and guidebook development.**

    (a) An android-based mobile application was developed to provide single-window platform to all stakeholders.

    (b) A guidebook for the interested consumers was developed.

(iii) **Site assessment.**

    (a) Rooftop measurement of interested consumers was undertaken.

(iv) **Capacity building.**

    (a) Capacity building workshops for industrial and residential consumers, as well as societies were conducted.

(v) **Installation.**

    (a) Around 15 MW of RTS was installed.

    (b) Around 11 MW of 15 MW was installed for the residential category.

2. Collab–Solar Project

The World Resources Institute (WRI) India and Confederation of Indian Industry (CII) through the Green Power Market Development Group initiative attempted to aggregate energy demand from six corporate buyers in Bengaluru to demonstrate a new aggregated procurement model that could be replicated in India to accelerate the deployment of RTS power.[7] The model was implemented in 2014 through the following key activities:

(i) consultation with potential buyers by

    (a) meeting with the potential buyers and

    (b) socializing the concept of the program with the buyers;

(ii) data collection and verification through

    (a) questionnaire-based survey conducted with the buyers and

    (b) preliminary site visits conducted by WRI India,

(iii) request for qualification (RFQ) processing, which included

    (a) RFQ preparation,

    (b) release into markets,

    (c) response analysis, and

    (d) results presentation to buyers;

(iv) capacity building through

    (a) webinars conducted by WRI India,

    (b) price and terms of negotiations,

    (c) due diligence,

    (d) analysis of responses, and

    (e) selection of vendors;

---

[7] A. K. Thanikonda, D. S. Krishnan, and S. Srivatsa. 2015. Aggregating Demand For Corporate Rooftop Solar Installations: Lessons from the Collaborative Solar PV Procurement Project. WRI India and CII Working Paper. November. https://gpmdg.org/case-studies/aggregating-demand-for-corporate-rooftop-solar-installations/.

    (v) request for proposal (RFP) processing, which included

        (a) RFP preparation,

        (b) release into market,

        (c) vendor site visits,

        (d) analysis of proposals, and

        (e) results presentation to buyers;

    (vi) negotiations and contracting, which included

        (a) price and terms of negotiations,

        (b) due diligence,

        (c) analysis of responses,

        (d) selection of the winning vendor, and

        (e) contract signing.

As key stakeholders, the

    (i) CII

        (a) organized two consultation meetings which were attended by large corporate buyers, clean energy suppliers, and officials from the Karnataka state and

        (b) released the RFQ and RFP into the market;

    (ii) WRI India

        (a) presented its experience with aggregating demand for solar photovoltaic in California and lessons learned from it,

        (b) prepared the questionnaire and canvassed it among the buyers,

        (c) conducted site visits to verify the data provided by the participants,

        (d) prepared and circulated RFQ documents and undertook response analysis,

        (e) conducted webinars and briefings for information sharing and capacity building to give all buyers a common point of reference,

        (f) prepared and circulated the RFP and carefully evaluated the same,

        (g) presented results to the buyers, and

        (h) anchored the contract signing process;

    (iii) vendors made visits to buyer sites before filling the RFP;

    (iv) buyers undertook due diligence, participated in vendor selection and contract signing.

It was observed that the best output was achieved from the aggregation with three to five large buyers having a minimum demand of 1 mega unit per annum along with several small buyers. Large buyers with decent credit ratings reduced the cost of financing and elicited greater confidence from the financial institution to invest in the project. Where buyers were located in close proximity, the transaction and logistic costs reduced even further.

3. Solarise Dwarka

The Solarise Dwarka initiative is being implemented by BSES Rajdhani Power Limited (BRPL) in collaboration with GIZ India under its Indo-German Solar Partnership project.[8] It is a "utility-anchored rooftop program" for the

---

[8]   BSES Rajdhani Power Limited. BRPL Solar City Initiative: India's 1st Utility Anchored Solar Rooftop Consumer Aggregation Program. Information Booklet. http://solarbses.com/viewFile.aspx?filename=Solar_booklet.pdf.

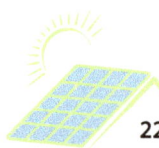

residential sector to maximize the utilization of RTS potential in south and west Delhi. As the anchor creating a platform to catalogue and aggregate all interested rooftop consumers, BRPL is

    (i)   bridging the gap between the consumer and the vendor,

    (ii) conducting post-installation inspection for assessing quality and ensuring adherence to the minimum technical requirements, and

    (iii) providing net metering facilities to the consumer.

As the key stakeholders,

    (i)   consumers

        (a) select the installer,

        (b) pay application and connectivity charges,

        (c) coordinate with the installer for timely installation, and

        (d) collect all documents for submission to the utility; while

    (ii) BRPL

        (a) publishes distribution capacity register,

        (b) provides information to the consumer for vendor selection,

        (c) conducts feasibility analysis at the application stage,

        (d) scrutinizes the documentation, and

        (e) tests and installs the solar meter.

Under this program, RTS installations are being provided at a single point for each apartment complex. The BRPL is aggregating all interested rooftop consumers in its distribution jurisdiction as well as the RTS vendors and integrators who are impaneled by IPGCL, Solar Energy Corporation of India, or the MNRE. The program is proposed to be run in line with the Delhi Electricity Regulatory Commission Net Metering Regulations, 2014 and guided by the Delhi Solar Policy, 2016 aiming to maximize the utilization of RTS potential in Delhi. In the first phase, only housing societies in Dwarka are being targeted; later, the program may be expanded to other areas.

4. Surya Urja Rooftop Yojana —Gujarat

The Government of Gujarat launched the "Surya Urja Rooftop Yojana (SURYA)—Gujarat" in August 2019, with a target of deploying enough RTS systems to cater to 200,000 consumers during FY2020 and a cumulative 800,000 consumers by FY2022.[9] Total aggregating capacity of RTS for 800,000 consumers is expected to be around 1,600 MW.[10] Under the program, consumer can install RTS above 1 kWp irrespective of the sanctioned load or contract demand. Subsidy is available from both the MNRE and the state up to 10 kWp capacity.

As the key stakeholders,

    (i)   consumers

        (a) select the installer,

        (b) pay application and connectivity charges,

        (c) coordinate with the installer for timely installation, and

        (d) collect all documents for submission to the utility; while

---

[9]   Government of Gujarat, Energy and Petrochemicals Department. 2019. Surya Urja Rooftop Yojana—Gujarat.

[10] Express News Service. 2019. Gujarat: Solar rooftop subsidy scheme targets 8 lakh homes in 3 yrs. 10 September. https://indianexpress.com/article/cities/ahmedabad/gujarat-solar-rooftop-subsidy-scheme-targets-8-lakh-homes-in-3-yrs-5980865/.

(ii) Gujarat Urja Vikas Nigam Limited (GUVNL)

   (a) impanels vendors for RTS installation,

   (b) ensures price discovery of various capacities of RTS,

   (c) collects expressions of interest from consumers,

   (d) provides information to consumers for vendor selection,

   (e) facilitates subsidy disbursement,

   (f) scrutinizes documents and monitors RTS installation, commissioning and O&M, and

   (g) tests and installs the solar meter.

Under this program, GUVNL, the nodal agency, launched a tender for 600 MW capacity adopting the CAPEX model. Through the tender, GUVNL impaneled more than 450 vendors for the consumers to select from for RTS installation on their roofs at a price discovered through the tendering process.

5. CREST—Demand Aggregation Initiative of the Center for Study of Science, Technology and Policy

The Center for Study of Science, Technology and Policy (CSTEP), Bengaluru, has been working with Bangalore Electricity Supply Company (BESCOM) for nearly a decade, applying the strategy framework in the Bengaluru area to help BESCOM achieve its 1.2 GW rooftop photovoltaic (RTPV) target for FY2022.

In a first-of-its-kind study in India, CSTEP used aerial light detection and ranging (LiDAR) to accurately map the potential of each rooftop in the city, after considering shading aspects and identify rooftops suitable for RTPV implementation. There are more than 870,000 rooftops in a 1,069 square kilometer area that have a cumulative capacity of 2.8 GW (Table 9). The database now has the locations of all of these rooftops along with the installable capacity. This is further categorized based on BESCOM sub-divisions. Iterations to this information in this phase of the proposed project will lead to the creation of a data room with the detailed project reports for each building. This can be used by vendors and BESCOM to aggregate demand to the tune of hundreds of megawatts, thereby achieving economies of scale for RTPV and driving down costs significantly. Vendors will also benefit in terms of reduced customer acquisition costs which will lead to further trimming of capital costs.

Table 9: Rooftop Photovoltaic Potential in Bengaluru Based on System Size

System size (in kilowatt)	Number of rooftop polygons	Cumulative capacity (in gigawatt)
> 50	4,513	0.5
10-50	40,953	0.8
< 10	825,253	1.5
	870,719	**2.8**

Source: Center for Study of Science, Technology and Policy, Bengaluru.

The LiDAR-based potential assessment led to the development of a free online tool, CREST, which shows accurate potential assessments and business cases considering consumption patterns. Any consumer in the surveyed area can use their BESCOM identification number (Account ID) to log into the tool and easily understand the technical and financial feasibility of installing RTPV. The tool uses BESCOM's database of consumption details of the user to calculate the business case and presents key findings including payback period and internal rate of return. The tool will now include the component of central government capital subsidy for residential consumers and highlight the magnified benefits of adopting RTPV. This will empower the user to make informed investment decisions and also earmark the shadow-free area on the roof as identified by CREST. This tool is being extensively used by BESCOM

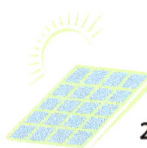

and developers to reach out to consumers. Future plans include promoting the tool through local nongovernment organizations and RTPV evangelists.

B. Investment Approach

1. Madhya Pradesh Urja Vikas Nigam Limited

The Madhya Pradesh Urja Vikas Nigam Limited (MPUVNL) invited bids under the Madhya Pradesh RESCO III. This was the first tender of RTS in India based on the demand aggregation model for the industries in Mandideep industrial area in Bhopal.

As key stakeholders, the
- (i) bidder
  - (a) undertook all works related to the commissioning and operations for 25 operational years of the project and
  - (b) checked whether the goods supplied were of the most recent technology, new and unused, and incorporated all recent improvements in design and materials as per the standards specified in the RFP;
- (ii) power producer
  - (a) synchronized the project with licensee's network under Madhya Pradesh Policy for Decentralized Renewable Energy Systems,
  - (b) ensured metering and grid connectivity of the projects,
  - (c) undertook O&M of the project up to a period of 25 years from the first part of commissioning or Scheduled Commercial Operation Date, whichever was earlier,
  - (d) purchased required meters,
  - (e) obtained insurance for third party liability covering loss of human life and risks of equipment after work completion, and
  - (f) oversaw the application of warranty and guarantee clauses and the resolution of claims and settlement of issues arising out of the said clauses;
- (iii) consumers
  - (a) signed PPA with the power producer and
  - (b) provided vacant roof space to the power producer for the implementation of project.

The developer was selected based on the reverse e-auction process. The capacity involved in this project was 10.8 MW. The discovered solar tariff of ₹4.61 per unit was less than MPUVNL's tariff of ₹7.5 per unit. About 164 (small, medium, and large) industries were to procure solar power under the RESCO model. The industries were put in three groups, each having a capacity of 3.6 MW. This project was implemented as part of the RTS program of the Government of India with the support from the World Bank.

2. Kerala State Electricity Board Soura Project

In 2018, with technical assistance from ADB, the Government of Kerala started its Soura scheme, to target the state's installation of ground-mounted and rooftop solar plants aggregating to 1,000 megawatt-peak (MWp). Kerala State Electricity Board (KSEB) created a Special Purpose Vehicle to smoothen the implementation of the project. Bids were invited to impanel EPC contractors and plant developers for design, supply, installation, testing, and commissioning of grid-tied rooftop and ground-mounted solar photovoltaic plants of the aggregated capacity of 50 MWp under the EPC category and 150 MWp under the tariff category, including O&M of the plant. Figure 10 represents key activities carried out from the launch of the Soura program to the selection of vendors.

Figure 10: Key Activities Carried Out by Kerala State Electricity Board

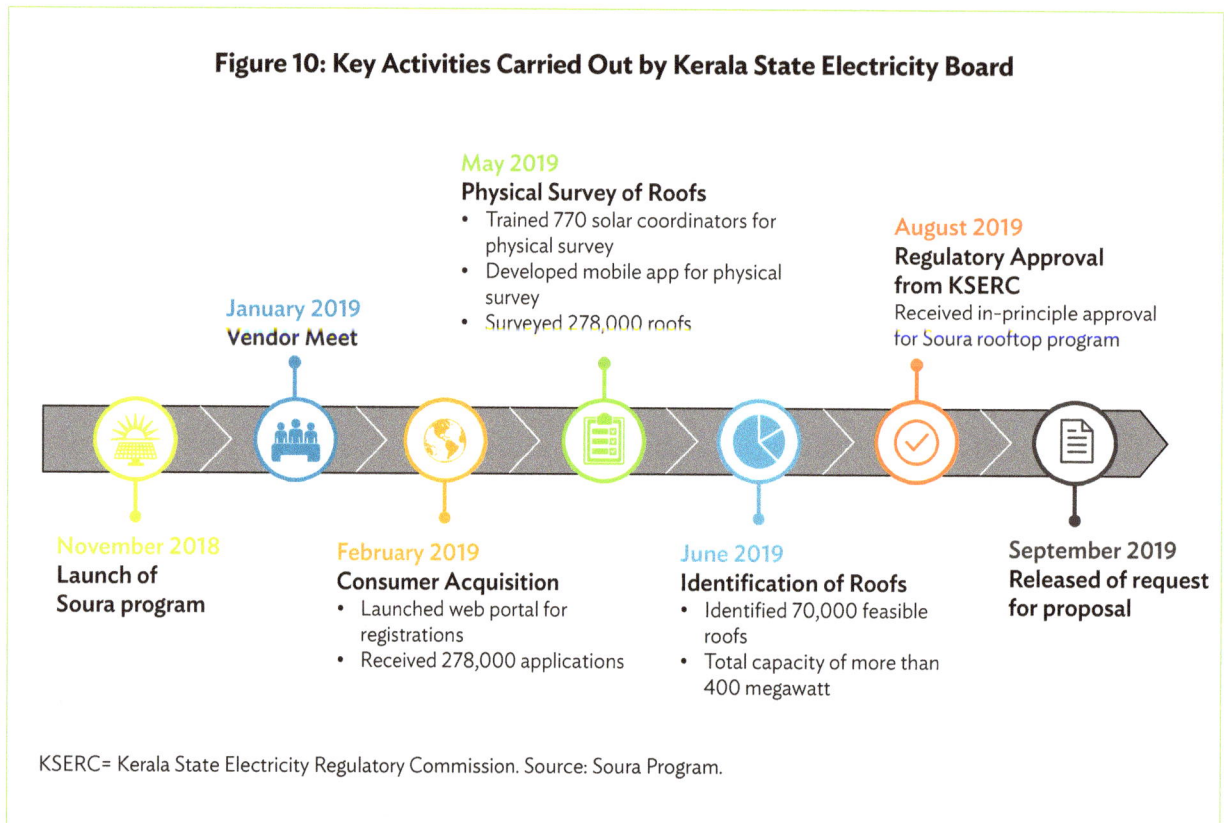

May 2019
Physical Survey of Roofs
- Trained 770 solar coordinators for physical survey
- Developed mobile app for physical survey
- Surveyed 278,000 roofs

August 2019
Regulatory Approval from KSERC
Received in-principle approval for Soura rooftop program

January 2019
Vendor Meet

November 2018
Launch of Soura program

February 2019
Consumer Acquisition
- Launched web portal for registrations
- Received 278,000 applications

June 2019
Identification of Roofs
- Identified 70,000 feasible roofs
- Total capacity of more than 400 megawatt

September 2019
Released of request for proposal

KSERC= Kerala State Electricity Regulatory Commission. Source: Soura Program.

As key stakeholders, the
- (i) EPC contractor (in the EPC mode)
    - (a) undertook installations and allied works up to the interconnection point as per the Central Electricity Authority Standards,
    - (b) installed net meters and solar energy meters,
    - (c) erected the distribution infrastructure from the solar generation facility to the interconnection point of equipment and data acquisition including communication facility,
    - (d) undertook O&M for the plant as per the mode selected,
    - (e) provided weather monitoring station at plant location for capacity above 50 kWp, and
    - (f) secured approval from the Electrical Inspectorate and made the connectivity application;
- (ii) state nodal officer (SNO)
    - (a) assisted in obtaining sanction, clearances, and approval from appropriate authorities,
    - (b) obtained consent of building- and landowners for installations,
    - (c) provided data room to be accessed by the contractors awarded the work,
    - (d) confirmed the quality and performance of RTS systems on behalf of the consumers, and
    - (e) routed payments to contractors, thereby assuring them appropriate and risk-free flow of payments;
- (iii) consumers
    - (a) submitted interest for the installation of the RTS project and selected the business model,
    - (b) signed agreement with SNO and KSEB based on the opted business model, and
    - (c) provided vacant roof space to the EPC contractor.

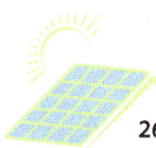

Online registration of the demand aggregation scheme was started by KSEB in July 2018 and closed on 31 January 2019. A total of 278,257 consumers are registered under three business models, namely,

(i) Roof lease model

(ii) Master RESCO model

(iii) EPC model

The KSEB conducted a field survey of the residential consumers and subsequently developed a mobile application to capture the survey details. Later, a web-based program was developed for selecting the best roof out of 278,264 registrations; around 70,000 applicants had the best rooftops for solar installations. The KSEB engaged three EPC developers through a competitive bidding process for implementing 46.5 MW RTS projects under the aforementioned business models. As of March 2022, around 15.627 MW of RTS capacity has been deployed.

6 PLANNING AND UNDERTAKING DEMAND AGGREGATION PROGRAM

Aggregation of interest from consumers, assessment of sites, and selection of developers for the deployment of projects are the key activities to be undertaken by the utility for the implementation of the program. The approach of the utilities may vary based on the roles and responsibilities undertaken.

Selection of approach for demand aggregation is based on parameters such as investment capabilities of the utility and the availability of personnel to operate and maintain the rooftop systems. The approach is also dependent on cost–benefit assessment based on the selected consumer category or region (Tables 10 and 11).

Table 10: Selection of Approach to Demand Aggregation

Cases	Pre-installation	Post-installation	Preferred approach and rationale
Low-paying consumer category	Highly cross-subsidized utility tariff along with state government subsidy in a few states	(i) Long-term reduction in cross subsidy and benefit to high-paying consumer category (ii) Higher benefits to utilities compared to the lower revenue from the category (iii) Saving in state government subsidy	(i) Facilitation approach along with financing and EMI collection. (ii) As low-paying consumers may not have access to initial capital investment, the utility may collaborate with the financial institution to provide financing to the low-paying consumer and thereafter collecting EMI, which reduces the risks of banks and enables easy access to project financing. (iii) Energy charges of low-paying consumers are less than the average cost of supply; hence, the facilitation approach is preferred.
High-paying consumer category	(i) Energy charges are high and consumers naturally move to RTS as it is cheaper than grid power. (ii) Limited adoption of RTS among commercial and industrial consumers due to lack of awareness.	Loss of revenue is high as the consumer-categories with high energy charges cross subsidize low-paying consumer categories.	(i) Investment approach (ii) Utilities, to retain their consumer base, invest and install RTS plants on the roofs of high-paying consumers. The energy generated is sold to the consumer at a tariff lower than current utility tariff thereby considerably limiting the revenue loss. (iii) RPO benefit accrues to the utility.

continued on next page

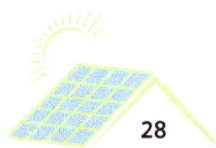

Table 10: continued

Cases	Pre-installation	Post-installation	Preferred approach and rationale
Utility invests as power generator	(i) Utility invests and install RTS on consumer's roof and procures the energy generated at no cost. Consumers receive rebate for the utilization of roof space. (ii) When rebate is calculated in terms of ₹ per kilowatt-hour, impact on utility revenue is the same, irrespective of consumer category. (iii) When rebate is calculated in terms of energy credits as a percentage of energy generated, revenue loss to utility is higher for high-paying consumer category.		(i) Utility invests as power generator. (ii) This approach is more beneficial to utilities where land is not available for large-scale deployment of solar photovoltaic plants. (iii) Utility operates and maintains RTS power plant. Utility can procure power at cheaper rates. (iv) Distribution losses are reduced because the power generated by the RTS plant is sold to the consumer at utility energy rates. (v) RPO benefits accrue to the utility.

EMI = equated monthly installment, RPO = renewable purchase obligation, RTS = rooftop solar.
Source: ADB Solar Rooftop Investment Program Technical Assistance.

Table 11: Parameters to Be Considered while Targeting Consumer Category or Region for Demand Aggregation

Parameter	Decision point
Gap between average revenue realization (ARR) and average cost of supply (ACOS)	(i) In case ARR > ACOS, the investment approach is preferred as shifting of consumers to rooftop solar (RTS) independently would lead to heavy revenue loss to the utility. These consumer categories cross subsidize low-paying consumer categories. (ii) For consumer categories, where ARR < ACOS, the facilitation approach can be used.
Billing efficiency	(i) In areas with lower billing efficiency, the facilitation approach is preferred to limit losses from low billing revenue. (ii) In areas with high billing efficiency, the investment approach is preferred so that the risk of cost–benefit is reduced.
Distribution losses	(i) Areas with high distribution losses are preferred for RTS deployment. (ii) Localized consumption from RTS systems reduces distribution losses and helps avoid costs of immediate infrastructure upgrade.
Power interruptions	Areas with minimum power interruptions are preferred to maximize the use of solar power generation.
Loading of distribution transformers	Distribution transformers with higher loading are preferred to reduce daytime loading.
Accessibility to site	Accessibility to site is important for the installation and maintenance of rooftop plants. Regular maintenance elicits best performance from rooftop plants.
Requirement of distribution system upgrade	Utilities prefer areas where there is no immediate need to upgrade infrastructure.

Source: ADB Solar Rooftop Investment Program Technical Assistance.

A. Key Steps to Demand Aggregation

Demand aggregation of RTS includes invitation of interest from consumers, assessment of roofs and installation, and O&M of RTS plant. The role of the utility and other stakeholders varies with program design and approach preferred

by the utility. Figure 11 depicts the key activities undertaken in an RTS demand aggregation program. The last stage of the program, i.e., bid process management and vendor engagement are applicable only to the investment approach.

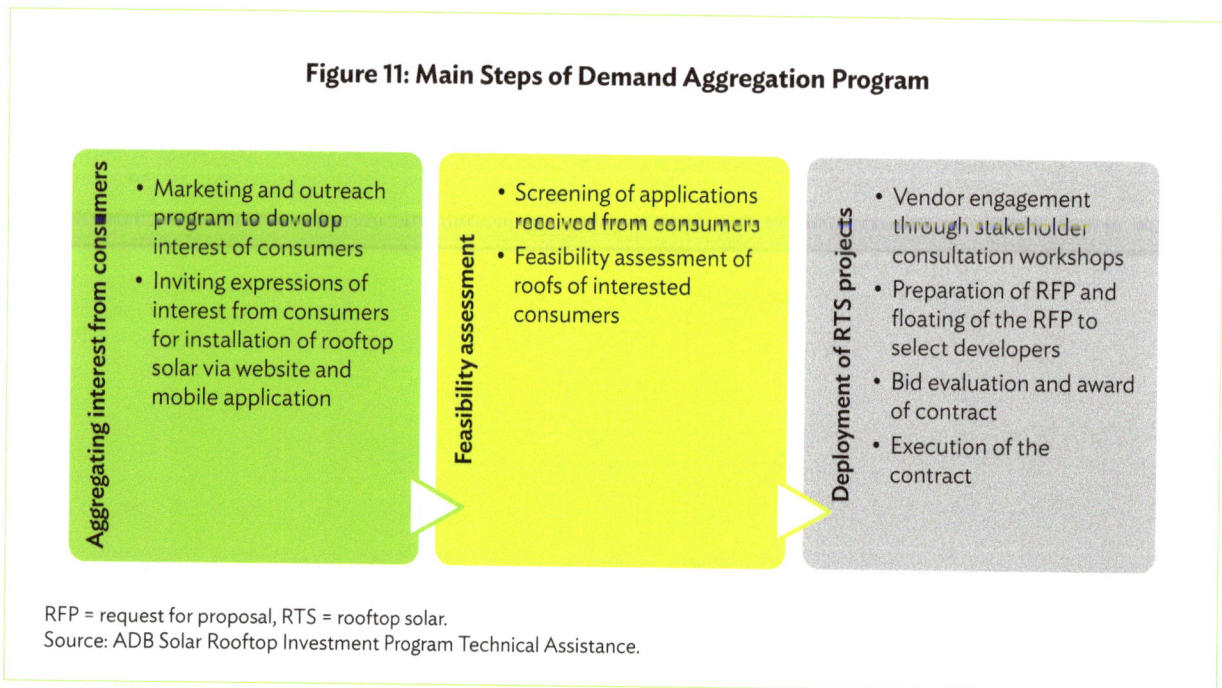

Figure 11: Main Steps of Demand Aggregation Program

Aggregating interest from consumers	Feasibility assessment	Deployment of RTS projects
• Marketing and outreach program to develop interest of consumers • Inviting expressions of interest from consumers for installation of rooftop solar via website and mobile application	• Screening of applications received from consumers • Feasibility assessment of roofs of interested consumers	• Vendor engagement through stakeholder consultation workshops • Preparation of RFP and floating of the RFP to select developers • Bid evaluation and award of contract • Execution of the contract

RFP = request for proposal, RTS = rooftop solar.
Source: ADB Solar Rooftop Investment Program Technical Assistance.

B. Planning Demand Aggregation Program

Large-scale RTS demand aggregation models require detailed planning on technical assessment, financial viability, and human resource requirement. Figure 12 displays the steps of demand aggregation program.

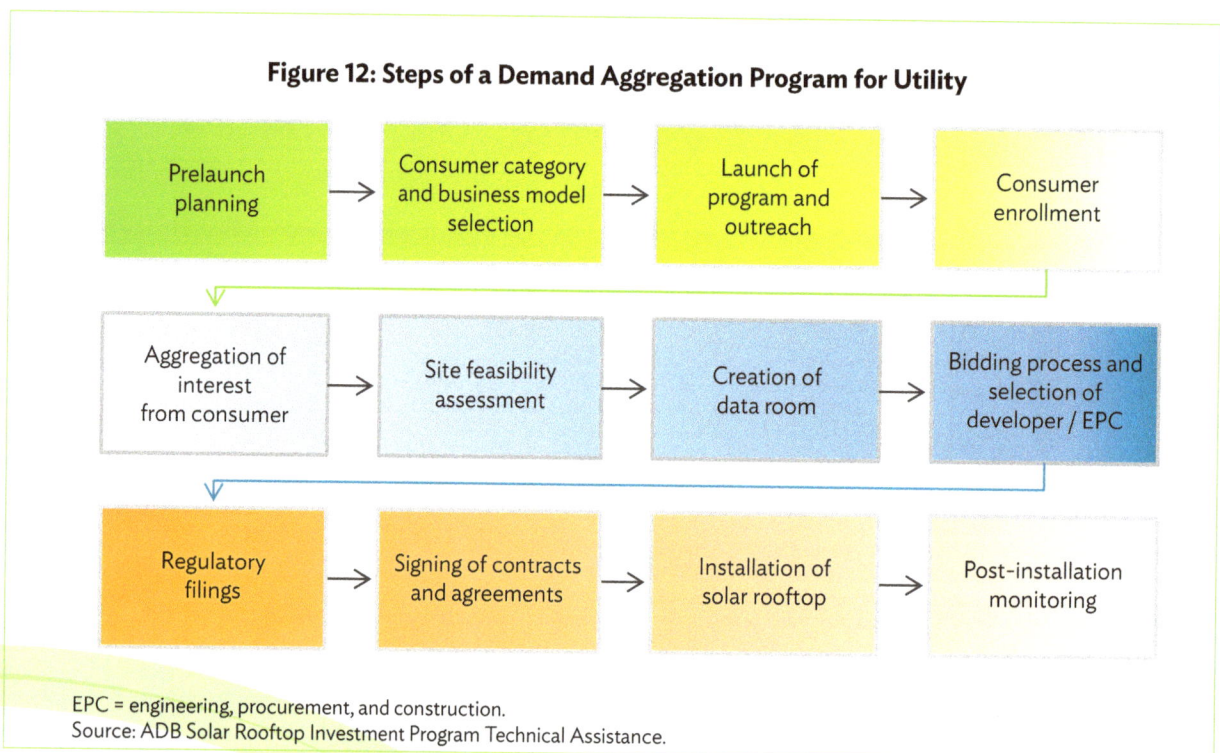

Figure 12: Steps of a Demand Aggregation Program for Utility

Prelaunch planning → Consumer category and business model selection → Launch of program and outreach → Consumer enrollment

Aggregation of interest from consumer → Site feasibility assessment → Creation of data room → Bidding process and selection of developer / EPC

Regulatory filings → Signing of contracts and agreements → Installation of solar rooftop → Post-installation monitoring

EPC = engineering, procurement, and construction.
Source: ADB Solar Rooftop Investment Program Technical Assistance.

C. Step 1: Prelaunch Planning

1. Part A: Capacity Development of Utility

The utility is an important stakeholder in facilitating the deployment of RTS in the state. Awareness and understanding of RTS are important right from senior management to field officials. Decision makers should have the knowledge of technical aspects as well as the competence to assess the financial viability of RTS systems. Field officials need to be trained in technical aspects of RTS as they are responsible for the physical assessment of roofs as well as inspection and installation of net meters on site.

Training programs should be customized as per the requirement of the utility and target group of officials. Table 12 illustrates the basic structure of the training programs in terms of technical, commercial, and regulatory aspects.

Table 12: Basic Structure of the Training Programs for Utility

Technical session	Commercial and regulatory session
Rooftop solar (RTS) system components	Interconnection arrangements
Balance of system components—quality and standards	Guidelines of the Ministry of New and Renewable Energy / Bureau of Indian Standards/Indian Standards Institution
Configuration of net and gross metered system	State regulations and policy
System capacity decision	Billing and energy accounting and availability of roof and fund
Energy generation estimation	Applicable standards for grid connectivity
Interconnection of RTS system	Interconnection requirements and technical specifications
Safety aspects of RTS and protection equipment	Energy accounting and commercial settlement for metering arrangements
Component inspection checklist	Levelized cost of energy calculation of RTS
Grid-connected functional safety checklist	
Commissioning, pre-commissioning, and post-commissioning tests	

Source: ADB Solar Rooftop Investment Program Technical Assistance.

A field visit to an RTS plant is recommended to create better understanding of the technology and provide hands-on experience to utility officials on technical and operational aspects of an RTS plant.

2. Part B: Formation of a Dedicated Team

A demand aggregation scheme is developed based on the demographics, consumer category, cost–benefit, and business model. Within a state regulatory framework, a business model can be developed considering the economics and the interests of stakeholders.

Demand aggregation of RTS involves a large number of stakeholders and buildings. To coordinate and monitor proper execution of the program, a three-tier team is envisaged—a technical committee, a head-office program team, and ground-level program team.

(i) The **technical committee** includes senior management of the utility, the state regulator, the chief electrical inspectorate, and the state renewable energy development agency (Table 13). A robust technical committee supports quick decision-making and improves ease of receiving approvals.

Table 13: Members of the Technical Committee

Proposed members of technical committee
Chief Engineer, Distribution
Chief Engineer, Technical
Chief Engineer, Commercial
State Nodal Officer, Rooftop solar photovoltaic project
Chief Electrical Inspector
Senior Official of State Nodal Agency
Representatives from Technical Assistance Programs of the Ministry of New and Renewable Energy, Government of India

Source: ADB Solar Rooftop Investment Program Technical Assistance.

The technical committee addresses technical issues, if any, during the pre-bid meeting, supports evaluation of technical bids for pre-qualification and the evaluation of price bids, advises the utility on award of contract, and oversees the role during the implementation of the RTS project.

(ii) The **head office program team** includes technical, commercial, and legal officials within the utility to execute, coordinate, and monitor the RTS demand aggregation program. The key roles of the team include receiving consumer enrollment, resolving consumer queries, assessing the physical feasibility of aggregated roofs, selection of developers, and monitoring the installation of RTS systems.

(iii) The **ground-level program team** includes divisional and subdivisional engineers of the utility who execute physical assessment of sites. The same team is primarily responsible for site inspection, verification, and monitoring timely installation of RTS projects.

D. Step 2: Consumer Category and Business Model Selection

Choice of consumer category and business model is important for defining the demand aggregation program.

Therefore, it is important for the utility to define the target consumer category and business model. To assess the cost–benefit of the demand aggregation program and to select the target consumer category, the following factors shall be considered:

(i) **Energy requirement and peak demand.** Short-term power purchase is generally very expensive compared to long-term power purchase. Adoption of RTS can reduce day-time energy requirement deficit. Deployment of RTS is comparatively beneficial in commercial and industrial consumers as they have high day time load, and localized consumption of power will reduce losses.

(ii) **Transmission and distribution losses.** While assessing the cost of supply, transmission and distribution (T&D) losses are added to average power purchase cost. With RTS, the consumer consumes power locally, thereby reducing T&D losses. Deployment of RTS also helps avoid immediate system upgrade expenses in areas where distribution losses are high.

(iii) **Power purchase cost assessment.**

   (a) **Marginal power purchase cost** is considered if the deployment of RTS plants substitute generators with high variable charges.

   (b) In a few states, where there are multiple utilities, utilities do not have the option to choose the power generator to procure power. State regulators combine all the generators and then based on the power

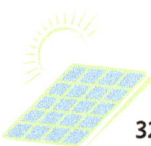

requirement, the generators are allotted to each utility. In such cases, where it is not possible to determine the power generators in future and estimate marginal power purchase cost, **average power purchase cost** is considered.

(iv) **Energy charges of targeted consumer category.** High-paying consumer categories are more inclined to installing RTS. Hence, the utility should seize the opportunity as an investor, deploy RTS for high-end consumers, and sell the energy generated from the RTS to the consumer. At the same time, the utility should be willing to support the development of RTS systems for subsidized consumer categories to improve its bottom line.

(v) **Revenue gap and subsidy provided by the state government.** Some states provide subsidy for the supply of power to specific consumer categories such as consumers below poverty line or agricultural consumers. Utilities may use the state subsidies fund for the deployment of RTS to these consumer categories and provide supply at no cost for the life of the project.

(vi) **Solar rooftop targets.** Targets of RTS set by the MNRE and state governments must be achieved by the utilities. Further, utilities that are unable to fulfill RPO and compelled to purchase renewable energy certificates (REC) could consider the benefit of RTS deployment in terms of savings in purchasing REC.

Consumer-category specific parameters to assess the development of RTS program include

(i) energy sales,

(ii) number of consumers,

(iii) connected load,

(iv) average billing rate and average revenue realization,

(v) average cost of supply, and

(vi) retail tariff.

E. Step 3: Launch of Program and Outreach

A major barrier to the deployment of RTS is the lack of awareness among consumers regarding the technology and project feasibility. Utilities could launch a media campaign for the promotion of the RTS program. An RTS program targeting a specific consumer category under a specific business model could be developed or a page or portal could be integrated in the utility's website with the provision for enrolling consumers. Outreach could be undertaken via social and print media. Volunteers could also be involved to spread awareness about the scheme.

A workshop could be organized by the utility for targeted consumers to improve their understanding of RTS technology and cost–benefit of installing RTS systems. The workshop could cover basics of technical assessment, technology, and financial benefits of adopting RTS.

F. Step 4: Consumer Enrollment

Utilities have existing relations with their consumers and have a ready database of basic information about the consumer such as address, connected load, etc. The utility aggregates demand for RTS from the consumers by receiving expressions of interest (EOIs) from the consumers either online or offline. By signing an EOI, the consumer agrees to participate in the installation of RTS and allows the utility to undertake feasibility assessment. The utility impanels developers and undertakes price discovery. A formal contract is signed between the consumer, utility, and developer based on the business model.

G. Step 5: Aggregation of Interest from Consumers

For large-scale demand aggregation at the district or state level, online mode (through a web portal or page) is preferred for inviting EOIs from individual consumers (whose basic information is already there with the utility). In cases where

demand aggregation is focused on industrial complexes or housing societies, offline EOIs may be considered. Under offline mode, field engineers of the utility go door-to-door collecting data and getting EOIs signed from the consumers. Table 14 depicts a template for offline data collection.

Table 14: Offline Data Collection Template

Consumer input	
Consumer number	(from utility electricity bill)
Mobile number of contact person	
E-mail ID of contact person	
Information fetched from distribution utility database	
Name of applicant	
Address of applicant	
Subdivision name and code	
Distribution transformer name and code (to be filled by the distribution company)	
Contract demand or connected load in kilowatt-peak (kWp)	
Connection type	1-phase low tension / 3-phase low tension/ high tension
Tariff consumer category	
Information collected from consumer	
Proposed rooftop capacity (kWp)	
Approximate shadow-free land area within the premises (square meter)	
Type of grid connectivity	Net metering or gross metering or net billing or virtual or group net metering
Latitude of the roof for installation	
Longitude of the roof for installation	
Choice of business models	Options of unique business models
Mode of implementation	Capital expenditure / Operational expenditure
Signature of consumer	

Source: ADB Solar Rooftop Investment Program Technical Assistance.

Online Expression of Interest from Consumers

Depending on the means of outreach preferred by the utility for a large-scale program, i.e., for subdivision or division or state, a website could provide a centralized platform for inviting interest. If the utility accepts the invitation of interest on a door-to-door basis for a selected area, i.e., for a housing society or industrial cluster, a mobile application may be developed on similar format (Figure 13).

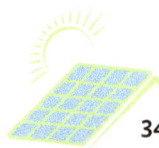

Figure 13: Format for Online Expression of Interest

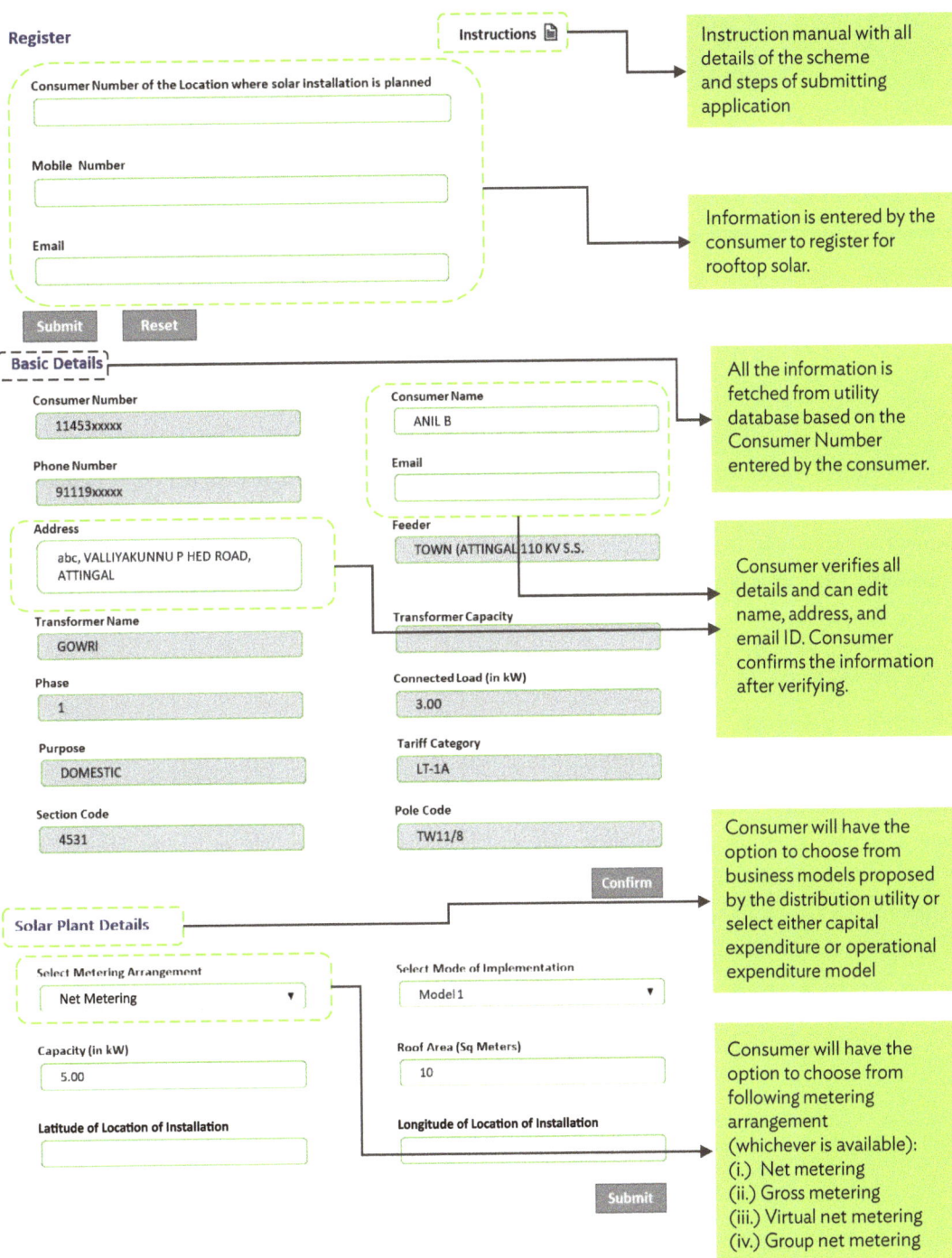

Register

Instructions 📄

Consumer Number of the Location where solar installation is planned

[   ]

Mobile Number

[   ]

Email

[   ]

[Submit] [Reset]

> Instruction manual with all details of the scheme and steps of submitting application

> Information is entered by the consumer to register for rooftop solar.

Basic Details

Consumer Number

11453xxxxx

Phone Number

91119xxxxx

Address

abc, VALLIYAKUNNU P HED ROAD, ATTINGAL

Transformer Name

GOWRI

Phase

1

Purpose

DOMESTIC

Section Code

4531

Consumer Name

ANIL B

Email

[   ]

Feeder

TOWN (ATTINGAL 110 KV S.S.

Transformer Capacity

[   ]

Connected Load (in kW)

3.00

Tariff Category

LT-1A

Pole Code

TW11/8

[Confirm]

> All the information is fetched from utility database based on the Consumer Number entered by the consumer.

> Consumer verifies all details and can edit name, address, and email ID. Consumer confirms the information after verifying.

Solar Plant Details

Select Metering Arrangement

Net Metering ▼

Capacity (in kW)

5.00

Latitude of Location of Installation

[   ]

Select Mode of Implementation

Model 1 ▼

Roof Area (Sq Meters)

10

Longitude of Location of Installation

[   ]

[Submit]

> Consumer will have the option to choose from business models proposed by the distribution utility or select either capital expenditure or operational expenditure model

> Consumer will have the option to choose from following metering arrangement (whichever is available):
> (i.) Net metering
> (ii.) Gross metering
> (iii.) Virtual net metering
> (iv.) Group net metering

Source: Soura Program.

All data collected would be verified, sorted, and compiled in a data room created for the purpose. The data room would consist of all the information required to understand the consumer profile, roof area, and feasible capacity. This data would be shared with the developers to facilitate more competitive price discovery for the implementation of projects.

H. Step 6: Site Feasibility Assessment

Feasibility assessment includes qualitative and quantitative assessment of roofs and the solar photovoltaic potential. Data collected in feasibility assessment of roofs provides preliminary information to the developers regarding the condition of the site, reduces risk for them, and assists them in quoting accurate price bids.

Feasibility assessment includes the following:

(i) **Access to site and access to roof.** Can the RTS plants be transported to the site and to the roofs?

(ii) **Configuration of roof.** Is the roof flat or slanted? In case of slanted roofs, the direction and angle of inclination of roof is to be measured.

(iii) **Material of roof.** Based on the type of material used for roofs, the mounting structure for photovoltaic modules is designed. Knowing the age of building helps in deciding the feasibility of the rooftop plant. Collecting the information on age and type of roof assists the developer in evaluating the exact cost of installation.

(iv) **Shadow-free roof area.** The roof area is measured and shadow area (based objects casting shadows such as walls and water tanks on the roof or trees and buildings adjacent to it) is subtracted to measure the shadow-free area.

(v) **Capacity of RTS to be installed.** Estimation of capacity is described below.

(vi) **Proximity to the interconnection point.** This provides information for the developer to estimate accurate cabling costs.

(vii) **Geo-coordinates of the site.** Though the same information is collected from the consumer during registration phase, it is verified by the official undertaking physical survey.

(viii) **Ownership of building.** This provides clarity on the ownership of the building and reduces risk for developers, especially in case on housing societies.

After the feasibility assessment, the RTS photovoltaic potential is assessed. Potential can be estimated by physical survey as well as simulation tools available online such as Helioscope, PVSyst, PVSol, and System Advisor Model. Primary determinants of the amount of energy delivered include

(i) rated capacity or size of the photovoltaic array (kWp),

(ii) amount of solar radiation it receives (expressed as peak sun hour), and

(iii) total efficiency of the system after considering all the losses.

(i) **Rated capacity or size of the photovoltaic array (kWp)**

An estimation of the potential installed capacity of the RTS system, P_{PV}, in kWp, may be derived using the following equation:

$$P_{PV} = (C_M/1000) \times (RCR \times A_R/A_M)$$

where,

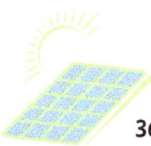

(a) A_R is the total encumbrance- and shadow-free roof area available for installation of solar modules in square meter (m^2),

(b) C_M is the individual module rated capacity in Wp,

(c) A_M is the area of one module in m^2, and

(d) RCR is the roof cover ratio, which is the fraction of roof area that the modules will cover. A typical value for the roof cover ratio (RCR) is 0.85, allowing 15% of the roof to be free for placing modules away from obstructions and maintaining spaces between them.

As a rule of thumb, approximate shadow-free roof area required for 1 kWp PV array is 10 m^2.

(ii) Amount of solar radiation received (expressed as peak sun hour)

Average global horizontal irradiance or peak sun hour is 4–6 kWh per m^2 in India. As it changes every day, only yearly average is considered.[11]

(iii) The total efficiency of the system after considering all the losses

Estimated loss for a RTS system depends on system design, component selection, and site operating temperature (Table 15). Normally, the total loss is around 26%.

Table 15: Typical Losses in Photovoltaic Systems

Cause of loss	Estimated losses (%)	Efficiency factor
Temperature	10	0.90
Dirt	3	0.97
Manufacturer's tolerance	3	0.97
Shading	2	0.98
Orientation	0	1.00
Tilt angle	1	0.99
Voltage drop	2	0.98
Invertors	5	0.95
Loss due to irradiance level	3	0.97
Total efficiency factor (multiplying all efficiency factors)		0.74

Note: Actual loss depends on site conditions.
Source: ADB Solar Rooftop Investment Program Technical Assistance.

Tools such as LiDAR, drones, and high resolution satellite images may also be utilized for the potential assessment study. However, it is important to ensure that the tool is accurate and allows the study to be completed within a stipulated time period and budget. We have included a case study on LiDAR-based potential assessment tool developed by the Center for Study of Science, Technology and Policy for the Bangalore Electricity Supply Company Limited.

---

[11] https://globalsolaratlas.info/map.

I. Step 7: Creation of Data Room

The data room consists of all the information collected from consumers and gathered through the feasibility assessment of the roofs (including feasibility reports of individual consumers). It is made accessible and available to potential developers in a transparent way for them to estimate costs and arrive at most competitive price quotations.

The Appendix presents a sample data room created for the Soura project, under which KSEB assessed the feasibility of more than 178,000 applications received for the installation of RTS plants.

J. Step 8: Bidding Process and Selection of Developers or Engineering, Procurement, and Construction Entities

To initiate the selection of developers for the deployment of RTS, a tender document is drafted according to the standards followed by the utilities. For implementation under the MNRE RTS scheme, the tender documents specify MNRE proposed conditions.

The tender documents include terms and conditions for the implementation of RTS, timeline for implementation, performance parameters, technical specifications of equipment used, and the bidding process.

Since a large number of buildings are involved in demand aggregation, the sites are categorized based on individual RTS plant capacity and region for the bidding process. Rooftop solar costs vary with the size of the plant. The MNRE also specifies different benchmark costs according to the size of the individual plant for fiscal year 2022 (Table 16).

Table 16: Rooftop Solar Benchmark Costs as Defined by the Ministry of New and Renewable Energy (in ₹ per kilowatt-peak)

Capacity	Special category states	General category states
1 kilowatt (kW)	56,210	51,100
Above 1 kW and up to 2 kW	51,670	46,980
Above 2 kW and up to 3 kW	50,330	45,760
Above 3 kW and up to 10 kW	49,100	44,640
Above 10 kW and up to 100 kW	45,800	41,640
Above 100 kW and up to 500 kW	42,980	39,080

Note: Special category states refer to North East Region including Sikkim, Himalayan states of Himachal Pradesh and Uttarakhand, and four union territories.

Source: ADB Solar Rooftop Investment Program Technical Assistance.

Therefore, it is suggested that buildings be categorized into groups based on RTS capacity. Further, where a large area such as an entire state is considered, the sites may be classified by subregion or district within the state and bid out accordingly.

After finalization and issuance of the tender, the utility should organize a bidders' meet to resolve queries regarding the tender document and present in detail the method used for demand aggregation, interest received from the consumers along with consumer profiles, data room, and terms and conditions of the tender document.

K. Step 9: Regulatory Filings (Applicable under Investment Approach)

Every state and Union Territory has its own regulations for RTS projects regarding consumer category, billing and energy settlement, and permissible metering arrangements.

When utilities enter the RTS business as investors, there are additional costs associated with the initiative, which have implications for the annual revenue requirement of the utilities. Therefore, the utilities must file a petition with the State Electricity Regulatory Commission (SERC) and Joint Electricity Regulatory Commission (JERC) for approval on the investment requirements under the following sections of the Electricity Act, 2003.

(i) *Section 86(1)(e) of the Electricity Act, 2003 empowers the Hon'ble Commission to promote electricity generation from renewable sources of energy by providing suitable measures for connectivity with the grid and sale of electricity to any person. The relevant section of Electricity Act, 2003 is extracted below:*

(ii) *"Section 86. (Functions of State Commission): - (1) The State Commission shall discharge the following functions, namely: [...]*

promote co-generation and generation of electricity from renewable sources of energy by providing suitable measures for connectivity with the grid and sale of electricity to any person, and also specify, for purchase of electricity from such sources, a percentage of the total consumption of electricity in the area of a distribution licensee;"

(iii) *Further, Section 86(1)(b) of the Electricity Act, 2003 empowers the Hon'ble Commission to regulate electricity purchase and procurement process of distribution licensees including the price at which electricity shall be procured from the generating companies or licensees or from other sources through agreements for purchase of power for distribution and supply within the state.*

(iv) *The relevant section of Electricity Act, 2003 is extracted below: "Section 86. (Functions of State Commission): - (1) The State Commission shall discharge the following functions, namely: - [...]*

(a) regulate electricity purchase and procurement process of distribution licensees including the price at which electricity shall be procured from the generating companies or licensees or from other sources through agreements for purchase of power for distribution and supply within the State;"

(v) Section 61(h) of the Electricity Act, 2003 mandates the promotion of co-generation and generation of electricity from renewable sources of energy. The relevant section of the Electricity Act, 2003 is extracted below:

(vi) "Section 61. (Tariff regulations): The Appropriate Commission shall, subject to the provisions of this Act, specify the terms and conditions for the determination of tariff, and in doing so, shall be guided by the following, namely:- [...]

(a) the promotion of co-generation and generation of electricity from renewable sources of energy;"

(vii) Section 62 of the Electricity Act, 2003 empowers the Hon'ble Commission to determine the tariff for supply of electricity by a generating company to a distribution licensee;

(viii) "Section 62. (Determination of tariff): - (1) The Appropriate Commission shall determine the tariff in accordance with the provisions of this Act for – (a) supply of electricity by a generating company to a distribution licensee:"

(ix) Section 63 of the Electricity Act, 2003 empowers the Hon'ble Commission to adopt the tariff if such tariff has been determined through transparent process of bidding.

Petition to the SERC and the JERC should include information on

(i) the need for RTS projects in the state

    (a) given the benefits and incentives available to utilities under the MNRE grid-connected RTS scheme Phase II,

    (b) especially in case state agencies are unable to fulfill the RPO targets and are compelled to buy the RECs to compensate,

    (c) based on targets defined by the National Solar Mission and state governments;

(ii) the proposed business model including

    (a) modalities of new business models such as

        i. target area or consumer category for which the business model is proposed,

        ii. current progress in the program such as estimated potential,

        iii. ceiling tariff for procurement of power from RTS plants, and

        iv. benefit to stakeholders and cost-benefit;

    (b) deviation from current regulations, if any, and

    (c) financing of the project;

(iii) the grounds for filing the petition.

L. Step 10: Signing of Contracts and Agreements (Applicable under both Facilitation and Investment Approach)

Under the facilitation approach, the utility only acts as a medium for RTS installation for the consumers. Under EPC mode of implementation, the developer signs an EPC contract with the consumer for the installation, commissioning, and O&M of the RTS plant, whereas under the RESCO mode of implementation, the developer signs a PPA with the consumer for the sale of power at a defined tariff.

Under the investment approach, the utility is responsible for the installation of the RTS plant. The developer signs a contract or agreement with the utility and all the transactions of the developer are performed through the utility. The consumer signs a "right to use" contract for the utility to use the roof for the installation and operation of the RTS plant. The consumer, in return, receives a lease rent or incentive.

M. Step 11: Installation of Rooftop Solar

Installation is undertaken by the selected developer within a specified time frame. Thereafter, the roofs are handed over to the utility or consumer, depending on the approach adopted.

Under the facilitation approach, the role of the utility is limited to supervising the installation, commissioning, and O&M contract. Under the investment approach, the utility monitors, operates, and maintains the RTS system for the life of the plant.

N. Step 12: Monitoring Post Installation (Applicable under the Investment Approach)

After the installation of the system, the developer should submit biannual (once every 6 months) reports to the utility on system performance and monthly energy generation.

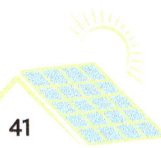

7   CONCLUSION AND WAY FORWARD

India has aimed for ambitious targets: to install additional capacities of 175 GW through renewable energy by 2022, and 500 GW by 2030. While the growth in solar capacity additions has been unprecedented, the RTS segment lags behind with only around 6.476 GW of capacity addition.

Rooftop solar offers many advantages to utilities, including reduced technical and commercial losses, better demand management, and savings in power procurement costs. Realizing these benefits, several utilities across various states have announced programs where they play an active role as RTS demand aggregators.

Through demand aggregation, capital cost and transaction cost can be reduced considerably. Consumers' limited awareness, and high consumer-acquisition costs for developers, have been major obstacles to the deployment of RTS. Both the issues can be addressed with the help of demand aggregation. As the photovoltaic market matures, RTS is expected to grow as it obviates the need for large land areas and reduces the load on the distribution grid owing to localized generation and consumption.

This guidebook will help utilities planning to undertake similar programs in their service areas to undertake appropriate demand aggregation for specific consumer categories. Utilities may use the guidebook to

(i)   create a training and capacity-building program at field level to equip front-facing officials on the benefits of RTS systems,

(ii) identify the approach to the demand aggregation program based on the parameters identified in the guidebook,

(iii) identify the target consumer category and select the appropriate business model,

(iv) invite interest from consumers and aggregate demand,

(v)   design a marketing campaign and market the program, and

(vi) launch the program and identify implementation modalities.

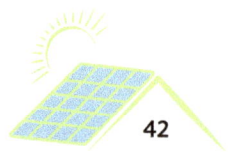

The screen shots presented in the Appendix are drawn from the data room of the SOURA Project Management App of the Kerala State Electricity Board Limited.

All ⦿   HT Only ⬭   ≤ 10 kwh ⬭   10 kwh to 100 kwh ⬭

Circles

Excel	PDF

Search: [   ]

Sl	Circle code	Circle name	Completed	Pending	Total
1	4021	Thiruvananthapuram Electrical Circle(Urban)	8074	0	8074
2	4022	Kattakada Electrical Circle (Rural)	4902	0	4902
3	4023	Kollam Electrical Circle	11496	0	11496
4	4024	Kottarakkara Electrical Circle	5211	0	5211
5	4025	Pathanamthitta Electrical Circle	7948	0	7948
6	4026	Kottayam Electrical Circle	8129	0	8129
7	5021	Alappuzha Electrical Circle	10838	26	10864
8	5022	Ernakulam Electrical Circle	8535	0	8535
9	5023	Perumbavoor Electrical Circle	17446	6	17452
10	5024	Thodupuzha Electrical Circle	11184	0	11184
11	5025	Pala Electrical Circle	5477	0	5477
12	5026	Irinjalakuda Electrical Circle	13586	0	13586
13	5027	Thrissur Electrical Circle	26577	0	26577
14	5028	Harippad Electrical Circle	6046	0	6046
15	6021	Palakkad Electrical Circle	15139	0	15139
16	6022	Manjeri Electrical Circle	9224	0	9224
17	6023	Kalpetta Electrical Circle	20233	560	20793
18	6024	Kozhikode Electrical Circle	10919	0	10919
19	6025	Kannur Electrical Circle	12081	0	12081
20	6026	Kasaragod Electrical Circle	16155	0	16155
21	6027	Tirur Electrical Circle	7445	0	7445
22	6028	Vadakara Electrical Circle	9944	0	9944
23	6029	Shoranur Electrical Circle	9645	0	9645
24	6030	Sreekandapuram Electrical Circle	10235	0	10235
25	6031	Nilambur Electrical Circle	9286	0	9286

Showing 1 to 25 of 25 entries

Previous 1 Next

Grades

Excel	PDF

Search: [   ]

Sl	Grade	Count	Capacity
1	A	42528	276868.00
2	B	62222	351558.18
3	C	60505	354762.00
4	D	2271	31511.00
5	E	108446	408606.50
	Total	275972	1423305.68

Showing 1 to 6 of 6 entries

Previous 1 Next

Circle: Thiruvananthapuram ElectricalCircle(Urban)

Excel   PDF   Search: _____

SI	Section code	Section name	Completed	Pending	Total
1	4506	Cantonment,TVM	157	0	157
2	4518	Nalanchira	279	0	279
3	4526	Edava	121	0	121
4	4504	Puthenchantha	51	0	51
5	4528	Kadakkavoor	97	0	97
6	4679	Poonthura	61	0	61
7	4507	Vellayambalam	205	0	205
8	4515	Sreevaraham	145	0	145
9	4680	Pallickal [Attingal Dvn]	130	0	130
10	4521	Kazhakuttam	408	0	408
11	4654	Nagaroor	107	0	107
12	4533	Kallambalam	242	0	242
13	4510	Karamana	145	0	145
14	4520	Kulathoor	165	0	165
15	4531	Attingal	214	0	214
16	4503	Fort,Trivandrum	54	0	54
17	4508	Peroorkada	315	0	315
18	4527	Kedakulam	150	0	150
19	4513	Beach,Trivandrum	149	0	149
20	4534	Palachira	195	0	195
21	4502	Thiruvallam	120	0	120
22	4678	Vattappara	97	0	97
23	4512	Poojappura	363	0	363
24	4519	Sreekariyam	447	0	447
25	4525	Varkala	194	0	194
26	4523	Mangalapuram	235	0	235
27	4529	Chirayinkeezhu	216	0	216
28	4509	Vattiyoorkavu	226	0	226
29	4517	Ulloor	238	0	238
30	4505	Thycaud	101	0	101
31	4691	Kachani	150	0	150
32	4522	Kaniyapuram	168	0	168
33	4535	Kilimannoor	162	0	162
34	4511	Thirumala	306	0	306

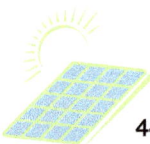

All ●   HT Only ◉   ≤ 10 kwh ◉   10 kwh to 100 kwh ◉

Section: Sreevaraham

Excel   PDF   Search: [   ]

Sl	Transformer	Completed	Pending	Total
1	KEEZHEVEEDU COLONY	14	0	14
2	CHEELANTHIMUKKU	12	0	12
3	MATHRUBHUMI	10	0	10
4	KANNIMEL KOTHALAM	10	0	10
5	SREENAGAR	7	0	7
6	VALIATHURA NO 1	6	0	6
7	FCI NO1	6	0	6
8	BEEMAPALLY NO 1	6	0	6
9	MUTTATHARA	5	0	5
10	ST ANNS	5	0	5
11	KUZHIVILAKOM	4	0	4
12	KOTHALAM	4	0	4
13	VALLAKADAVU	4	0	4
14	J P NAGAR	4	0	4
15	ENCHACKAL	4	0	4
16	KOOBAKKARA	3	0	3
17	VOR	3	0	3
18	VALIATHURA NO 2	3	0	3
19	THONDUKADAVU	3	0	3
20	ROSMINI CONVENT	3	0	3
21	ASHAN NAGAR	3	0	3
22	PERUNELLY	2	0	2
23	NS DEPOT	2	0	2
24	VELANKANNI	2	0	2
25	VARAHAM GARDEN	2	0	2
26	SIDHA	2	0	2
27	DEVI NAGAR	2	0	2
28	SEWAGE FARM	2	0	2
29	PRAMOG HT LT OWNER TR	2	0	2
30	BANGLADESH NO 2	2	0	2

Data Compiled for Individual Consumer

The data compiled for the individual consumer during the feasibility assessment shall be captured and available along with the utility data.

GEO-COORDINATES OF THE BUILDING, LAND

INDICATIVE SHADE FREE AREA

ROAD ACCESS

DETAILS OF THE BUILDING

PROXIMITY TO KSEB LTD. NETWORK

IMAGE OF THE BUILDING

ROOFTOP MEASUREMENTS WHEREVER POSSIBLE

The following figures depict the available information for each consumer concerning the feasibility of roof.

www.ingramcontent.com/pod-product-compliance
Lightning Source LLC
Chambersburg PA
CBHW041121280326

41928CB00061B/3484